CAMBRIDGE LIBRARY COLLECTION

Books of enduring scholarly value

Mathematical Sciences

From its pre-historic roots in simple counting to the algorithms powering modern desktop computers, from the genius of Archimedes to the genius of Einstein, advances in mathematical understanding and numerical techniques have been directly responsible for creating the modern world as we know it. This series will provide a library of the most influential publications and writers on mathematics in its broadest sense. As such, it will show not only the deep roots from which modern science and technology have grown, but also the astonishing breadth of application of mathematical techniques in the humanities and social sciences, and in everyday life.

Elementary Principles in Statistical Mechanics

Josiah Willard Gibbs (1839–1903) was the greatest American mathematician and physicist of the nineteenth century. He played a key role in the development of vector analysis (his book on this topic is also reissued in this series), but his deepest work was in the development of thermodynamics and statistical physics. This book, *Elementary Principles in Statistical Mechanics*, first published in 1902, gives his mature vision of these subjects. Mathematicians, physicists and engineers familiar with such things as Gibbs entropy, Gibbs inequality and the Gibbs distribution will find them here discussed in Gibbs' own words.

Cambridge University Press has long been a pioneer in the reissuing of out-of-print titles from its own backlist, producing digital reprints of books that are still sought after by scholars and students but could not be reprinted economically using traditional technology. The Cambridge Library Collection extends this activity to a wider range of books which are still of importance to researchers and professionals, either for the source material they contain, or as landmarks in the history of their academic discipline.

Drawing from the world-renowned collections in the Cambridge University Library, and guided by the advice of experts in each subject area, Cambridge University Press is using state-of-the-art scanning machines in its own Printing House to capture the content of each book selected for inclusion. The files are processed to give a consistently clear, crisp image, and the books finished to the high quality standard for which the Press is recognised around the world. The latest print-on-demand technology ensures that the books will remain available indefinitely, and that orders for single or multiple copies can quickly be supplied.

The Cambridge Library Collection will bring back to life books of enduring scholarly value (including out-of-copyright works originally issued by other publishers) across a wide range of disciplines in the humanities and social sciences and in science and technology.

Elementary Principles in Statistical Mechanics

Developed with Especial Reference to the Rational Foundation of Thermodynamics

Josiah Willard Gibbs

CAMBRIDGE UNIVERSITY PRESS

Cambridge, New York, Melbourne, Madrid, Cape Town, Singapore, São Paolo, Delhi, Dubai, Tokyo

Published in the United States of America by Cambridge University Press, New York

www.cambridge.org
Information on this title: www.cambridge.org/9781108017022

This edition first published 1902
This digitally printed version 2010

ISBN 978-1-108-01702-2 Paperback

Yale Bicentennial Publications

ELEMENTARY PRINCIPLES IN STATISTICAL MECHANICS

Yale Bicentennial Publications

With the approval of the President and Fellows of Yale University, a series of volumes has been prepared by a number of the Professors and Instructors, to be issued in connection with the Bicentennial Anniversary, as a partial indication of the character of the studies in which the University teachers are engaged.

This series of volumes is respectfully dedicated to

The Graduates of the University

ELEMENTARY PRINCIPLES

IN

STATISTICAL MECHANICS

DEVELOPED WITH ESPECIAL REFERENCE TO

THE RATIONAL FOUNDATION OF THERMODYNAMICS

BY

J. WILLARD GIBBS
Professor of Mathematical Physics in Yale University

NEW YORK: CHARLES SCRIBNER'S SONS
LONDON: EDWARD ARNOLD
1902

Published, March, 1902.

PREFACE.

The usual point of view in the study of mechanics is that where the attention is mainly directed to the changes which take place in the course of time in a given system. The principal problem is the determination of the condition of the system with respect to configuration and velocities at any required time, when its condition in these respects has been given for some one time, and the fundamental equations are those which express the changes continually taking place in the system. Inquiries of this kind are often simplified by taking into consideration conditions of the system other than those through which it actually passes or is supposed to pass, but our attention is not usually carried beyond conditions differing infinitesimally from those which are regarded as actual.

For some purposes, however, it is desirable to take a broader view of the subject. We may imagine a great number of systems of the same nature, but differing in the configurations and velocities which they have at a given instant, and differing not merely infinitesimally, but it may be so as to embrace every conceivable combination of configuration and velocities. And here we may set the problem, not to follow a particular system through its succession of configurations, but to determine how the whole number of systems will be distributed among the various conceivable configurations and velocities at any required time, when the distribution has been given for some one time. The fundamental equation for this inquiry is that which gives the rate of change of the number of systems which fall within any infinitesimal limits of configuration and velocity.

Such inquiries have been called by Maxwell *statistical.* They belong to a branch of mechanics which owes its origin to the desire to explain the laws of thermodynamics on mechanical principles, and of which Clausius, Maxwell, and Boltzmann are to be regarded as the principal founders. The first inquiries in this field were indeed somewhat narrower in their scope than that which has been mentioned, being applied to the particles of a system, rather than to independent systems. Statistical inquiries were next directed to the phases (or conditions with respect to configuration and velocity) which succeed one another in a given system in the course of time. The explicit consideration of a great number of systems and their distribution in phase, and of the permanence or alteration of this distribution in the course of time is perhaps first found in Boltzmann's paper on the "Zusammenhang zwischen den Sätzen über das Verhalten mehratomiger Gasmoleküle mit Jacobi's Princip des letzten Multiplicators" (1871).

But although, as a matter of history, statistical mechanics owes its origin to investigations in thermodynamics, it seems eminently worthy of an independent development, both on account of the elegance and simplicity of its principles, and because it yields new results and places old truths in a new light in departments quite outside of thermodynamics. Moreover, the separate study of this branch of mechanics seems to afford the best foundation for the study of rational thermodynamics and molecular mechanics.

The laws of thermodynamics, as empirically determined, express the approximate and probable behavior of systems of a great number of particles, or, more precisely, they express the laws of mechanics for such systems as they appear to beings who have not the fineness of perception to enable them to appreciate quantities of the order of magnitude of those which relate to single particles, and who cannot repeat their experiments often enough to obtain any but the most probable results. The laws of statistical mechanics apply to conservative systems of any number of degrees of freedom, and are exact. This does not make them more difficult to establish than the approximate laws for systems of a great many degrees of freedom, or for limited classes of such systems. The reverse is rather the case, for our attention is not diverted from what is essential by the peculiarities of the system considered, and we are not obliged to satisfy ourselves that the effect of the quantities and circumstances neglected will be negligible in the result. The laws of thermodynamics may be easily obtained from the principles of statistical mechanics, of which they are the incomplete expression, but they make a somewhat blind guide in our search for those laws. This is perhaps the principal cause of the slow progress of rational thermodynamics, as contrasted with the rapid deduction of the consequences of its laws as empirically established. To this must be added that the rational foundation of thermodynamics lay in a branch of mechanics of which the fundamental notions and principles, and the characteristic operations, were alike unfamiliar to students of mechanics.

We may therefore confidently believe that nothing will more conduce to the clear apprehension of the relation of thermodynamics to rational mechanics, and to the interpretation of observed phenomena with reference to their evidence respecting the molecular constitution of bodies, than the study of the fundamental notions and principles of that department of mechanics to which thermodynamics is especially related.

Moreover, we avoid the gravest difficulties when, giving up the attempt to frame hypotheses concerning the constitution of material bodies, we pursue statistical inquiries as a branch of rational mechanics. In the present state of science, it seems hardly possible to frame a dynamic theory of molecular action which shall embrace the phenomena of thermodynamics, of radiation, and of the electrical manifestations which accompany the union of atoms. Yet any theory is obviously inadequate which does not take account of all these phenomena. Even if we confine our attention to the phenomena distinctively thermodynamic, we do not escape difficulties in as simple a matter as the number of degrees of freedom of a diatomic gas. It is well known that while theory would assign to the gas six degrees of freedom per molecule, in our experiments on specific heat we cannot account for more than five. Certainly, one is building on an insecure foundation, who rests his work on hypotheses concerning the constitution of matter.

Difficulties of this kind have deterred the author from attempting to explain the mysteries of nature, and have forced him to be contented with the more modest aim of deducing some of the more obvious propositions relating to the statistical branch of mechanics. Here, there can be no mistake in regard to the agreement of the hypotheses with the facts of nature, for nothing is assumed in that respect. The only error into which one can fall, is the want of agreement between the premises and the conclusions, and this, with care, one may hope, in the main, to avoid.

The matter of the present volume consists in large measure of results which have been obtained by the investigators mentioned above, although the point of view and the arrangement may be different. These results, given to the public one by one in the order of their discovery, have necessarily, in their original presentation, not been arranged in the most logical manner.

In the first chapter we consider the general problem which has been mentioned, and find what may be called the fundamental equation of statistical mechanics. A particular case of this equation will give the condition of statistical equilibrium, *i. e.*, the condition which the distribution of the systems in phase must satisfy in order that the distribution shall be permanent. In the general case, the fundamental equation admits an integration, which gives a principle which may be variously expressed, according to the point of view from which it is regarded, as the conservation of density-in-phase, or of extension-in-phase, or of probability of phase.

In the second chapter, we apply this principle of conservation of probability of phase to the theory of errors in the calculated phases of a system, when the determination of the arbitrary constants of the integral equations are subject to error. In this application, we do not go beyond the usual approximations. In other words, we combine the principle of conservation of probability of phase, which is exact, with those approximate relations, which it is customary to assume in the "theory of errors."

In the third chapter we apply the principle of conservation of extension-in-phase to the integration of the differential equations of motion. This gives Jacobi's "last multiplier," as has been shown by Boltzmann.

In the fourth and following chapters we return to the consideration of statistical equilibrium, and confine our attention to conservative systems. We consider especially ensembles of systems in which the index (or logarithm) of probability of phase is a linear function of the energy. This distribution, on account of its unique importance in the theory of statistical equilibrium, I have ventured to call *canonical*, and the divisor of the energy, the *modulus* of distribution. The moduli of ensembles have properties analogous to temperature, in that equality of the moduli is a condition of equilibrium with respect to exchange of energy, when such exchange is made possible.

We find a differential equation relating to average values in the ensemble which is identical in form with the fundamental differential equation of thermodynamics, the average index of probability of phase, with change of sign, corresponding to entropy, and the modulus to temperature.

For the average square of the anomalies of the energy, we find an expression which vanishes in comparison with the square of the average energy, when the number of degrees of freedom is indefinitely increased. An ensemble of systems in which the number of degrees of freedom is of the same order of magnitude as the number of molecules in the bodies with which we experiment, if distributed canonically, would therefore appear to human observation as an ensemble of systems in which all have the same energy.

We meet with other quantities, in the development of the subject, which, when the number of degrees of freedom is very great, coincide sensibly with the modulus, and with the average index of probability, taken negatively, in a canonical ensemble, and which, therefore, may also be regarded as corresponding to temperature and entropy. The correspondence is however imperfect, when the number of degrees of freedom is not very great, and there is nothing to recommend these quantities except that in definition they may be regarded as more simple than those which have been mentioned. In Chapter XIV, this subject of thermodynamic analogies is discussed somewhat at length.

Finally, in Chapter XV, we consider the modification of the preceding results which is necessary when we consider systems composed of a number of entirely similar particles, or, it may be, of a number of particles of several kinds, all of each kind being entirely similar to each other, and when one of the variations to be considered is that of the numbers of the particles of the various kinds which are contained in a system. This supposition would naturally have been introduced earlier, if our object had been simply the expression of the laws of nature. It seemed desirable, however, to separate sharply the purely thermodynamic laws from those special modifications which belong rather to the theory of the properties of matter.

J. W. G.

NEW HAVEN, December, 1901.

CONTENTS.

CHAPTER I.

GENERAL NOTIONS. THE PRINCIPLE OF CONSERVATION OF EXTENSION-IN-PHASE.

CHAPTER II.

APPLICATION OF THE PRINCIPLE OF CONSERVATION OF EXTENSION-IN-PHASE TO THE THEORY OF ERRORS.

CHAPTER III.

APPLICATION OF THE PRINCIPLE OF CONSERVATION OF EXTENSION-IN-PHASE TO THE INTEGRATION OF THE DIFFERENTIAL EQUATIONS OF MOTION.

CHAPTER IV.

ON THE DISTRIBUTION-IN-PHASE CALLED CANONICAL, IN WHICH THE INDEX OF PROBABILITY IS A LINEAR FUNCTION OF THE ENERGY.

CHAPTER V.

AVERAGE VALUES IN A CANONICAL ENSEMBLE OF SYSTEMS.

CHAPTER VI.

EXTENSION-IN-CONFIGURATION AND EXTENSION-IN-VELOCITY.

CHAPTER VII.

FARTHER DISCUSSION OF AVERAGES IN A CANONICAL ENSEMBLE OF SYSTEMS.

CHAPTER VIII.

ON CERTAIN IMPORTANT FUNCTIONS OF THE ENERGIES OF A SYSTEM.

CHAPTER IX.

THE FUNCTION ϕ AND THE CANONICAL DISTRIBUTION.

CHAPTER X.

ON A DISTRIBUTION IN PHASE CALLED MICROCANONICAL IN WHICH ALL THE SYSTEMS HAVE THE SAME ENERGY.

CHAPTER XI.

MAXIMUM AND MINIMUM PROPERTIES OF VARIOUS DISTRIBUTIONS IN PHASE.

CHAPTER XII.

ON THE MOTION OF SYSTEMS AND ENSEMBLES OF SYSTEMS THROUGH LONG PERIODS OF TIME.

CHAPTER XIII.

EFFECT OF VARIOUS PROCESSES ON AN ENSEMBLE OF SYSTEMS.

CHAPTER XIV.

DISCUSSION OF THERMODYNAMIC ANALOGIES.

CHAPTER XV.

SYSTEMS COMPOSED OF MOLECULES.

ELEMENTARY PRINCIPLES IN STATISTICAL MECHANICS

ELEMENTARY PRINCIPLES IN STATISTICAL MECHANICS

CHAPTER I.

GENERAL NOTIONS. THE PRINCIPLE OF CONSERVATION OF EXTENSION-IN-PHASE.

WE shall use Hamilton's form of the equations of motion for a system of n degrees of freedom, writing $q_1, \ldots q_n$ for the (generalized) coördinates, $\dot{q}_1, \ldots \dot{q}_n$ for the (generalized) velocities, and

$$F_1\, dq_1 + F_2\, dq_2 \ldots + F_n\, dq_n \tag{1}$$

for the moment of the forces. We shall call the quantities $F_1, \ldots F_n$ the (generalized) forces, and the quantities $p_1 \ldots p_n$, defined by the equations

$$p_1 = \frac{d\epsilon_p}{d\dot{q}_1}, \quad p_2 = \frac{d\epsilon_p}{d\dot{q}_2}, \quad \text{etc.,} \tag{2}$$

where ϵ_p denotes the kinetic energy of the system, the (generalized) momenta. The kinetic energy is here regarded as a function of the velocities and coördinates. We shall usually regard it as a function of the momenta and coördinates,* and on this account we denote it by ϵ_p. This will not prevent us from occasionally using formulæ like (2), where it is sufficiently evident the kinetic energy is regarded as function of the $\dot{q}$'s and q's. But in expressions like $d\epsilon_p/dq_1$, where the denominator does not determine the question, the kinetic

* The use of the momenta instead of the velocities as independent variables is the characteristic of Hamilton's method which gives his equations of motion their remarkable degree of simplicity. We shall find that the fundamental notions of statistical mechanics are most easily defined, and are expressed in the most simple form, when the momenta with the coördinates are used to describe the state of a system.

energy is always to be treated in the differentiation as function of the p's and q's.

We have then

$$\dot{q}_1 = \frac{d\epsilon_p}{dp_1}, \quad \dot{p}_1 = -\frac{d\epsilon_p}{dq_1} + F_1, \quad \text{etc.} \tag{3}$$

These equations will hold for any forces whatever. If the forces are conservative, in other words, if the expression (1) is an exact differential, we may set

$$F_1 = -\frac{d\epsilon_q}{dq_1}, \quad F_2 = -\frac{d\epsilon_q}{dq_2}, \quad \text{etc.,} \tag{4}$$

where ϵ_q is a function of the coördinates which we shall call the potential energy of the system. If we write ϵ for the total energy, we shall have

$$\epsilon = \epsilon_p + \epsilon_q, \tag{5}$$

and equations (3) may be written

$$\dot{q}_1 = \frac{d\epsilon}{dp_1}, \quad \dot{p}_1 = -\frac{d\epsilon}{dq_1}, \quad \text{etc.} \tag{6}$$

The potential energy (ϵ_q) may depend on other variables beside the coördinates $q_1 \ldots q_n$. We shall often suppose it to depend in part on coördinates of external bodies, which we shall denote by a_1, a_2, etc. We shall then have for the complete value of the differential of the potential energy *

$$d\epsilon_q = -F_1\, dq_1 \ldots - F_n\, dq_n - A_1\, da_1 - A_2\, da_2 - \text{etc.,} \tag{7}$$

where A_1, A_2, etc., represent forces (in the generalized sense) exerted by the system on external bodies. For the total energy (ϵ) we shall have

$$d\epsilon = \dot{q}_1\, dp_1 \ldots + \dot{q}_n\, dp_n - \dot{p}_1\, dq_1 \ldots \\ - \dot{p}_n\, dq_n - A_1\, da_1 - A_2\, da_2 - \text{etc.} \tag{8}$$

It will be observed that the kinetic energy (ϵ_p) in the most general case is a quadratic function of the p's (or $\dot{q}$'s)

* It will be observed, that although we call ϵ_q the potential energy of the system which we are considering, it is really so defined as to include that energy which might be described as mutual to that system and external bodies.

involving also the q's but not the a's; that the potential energy, when it exists, is function of the q's and a's; and that the total energy, when it exists, is function of the p's (or $\dot{q}$'s), the q's, and the a's. In expressions like $d\epsilon / dq_1$, the p's, and not the $\dot{q}$'s, are to be taken as independent variables, as has already been stated with respect to the kinetic energy.

Let us imagine a great number of independent systems, identical in nature, but differing in phase, that is, in their condition with respect to configuration and velocity. The forces are supposed to be determined for every system by the same law, being functions of the coördinates of the system $q_1, \ldots q_n$, either alone or with the coördinates a_1, a_2, etc. of certain external bodies. It is not necessary that they should be derivable from a force-function. The external coördinates a_1, a_2, etc. may vary with the time, but at any given time have fixed values. In this they differ from the internal coördinates $q_1, \ldots q_n$, which at the same time have different values in the different systems considered.

Let us especially consider the number of systems which at a given instant fall within specified limits of phase, viz., those for which

$$\left.\begin{array}{ll} p_1' < p_1 < p_1'', & q_1' < q_1 < q_1'', \\ p_2' < p_2 < p_2'', & q_2' < q_2 < q_2'', \\ \cdots\cdots & \cdots\cdots \\ p_n' < p_n < p_n'', & q_n' < q_n < q_n'', \end{array}\right\} \qquad (9)$$

the accented letters denoting constants. We shall suppose the differences $p_1'' - p_1'$, $q_1'' - q_1'$, etc. to be infinitesimal, and that the systems are distributed in phase in some continuous manner,* so that the number having phases within the limits specified may be represented by

$$D\,(p_1'' - p_1') \ldots (p_n'' - p_n')\,(q_1'' - q_1') \ldots (q_n'' - q_n'), \qquad (10)$$

* In strictness, a finite number of systems cannot be distributed continuously in phase. But by increasing indefinitely the number of systems, we may approximate to a continuous law of distribution, such as is here described. To avoid tedious circumlocution, language like the above may be allowed, although wanting in precision of expression, when the sense in which it is to be taken appears sufficiently clear.

or more briefly by

$$D\,dp_1 \ldots dp_n\,dq_1 \ldots dq_n, \tag{11}$$

where D is a function of the p's and q's and in general of t also, for as time goes on, and the individual systems change their phases, the distribution of the ensemble in phase will in general vary. In special cases, the distribution in phase will remain unchanged. These are cases of *statistical equilibrium.*

If we regard all possible phases as forming a sort of extension of $2\,n$ dimensions, we may regard the product of differentials in (11) as expressing an element of this extension, and D as expressing the density of the systems in that element. We shall call the product

$$dp_1 \ldots dp_n\,dq_1 \ldots dq_n \tag{12}$$

an element of *extension-in-phase*, and D the *density-in-phase* of the systems.

It is evident that the changes which take place in the density of the systems in any given element of extension-in-phase will depend on the dynamical nature of the systems and their distribution in phase at the time considered.

In the case of conservative systems, with which we shall be principally concerned, their dynamical nature is completely determined by the function which expresses the energy (ϵ) in terms of the p's, q's, and a's (a function supposed identical for all the systems); in the more general case which we are considering, the dynamical nature of the systems is determined by the functions which express the kinetic energy (ϵ_p) in terms of the p's and q's, and the forces in terms of the q's and a's. The distribution in phase is expressed for the time considered by D as function of the p's and q's. To find the value of dD/dt for the specified element of extension-in-phase, we observe that the number of systems within the limits can only be varied by systems passing the limits, which may take place in $4\,n$ different ways, viz., by the p_1 of a system passing the limit p_1', or the limit p_1'', or by the q_1 of a system passing the limit q_1', or the limit q_1'', etc. Let us consider these cases separately.

In the first place, let us consider the number of systems which in the time dt pass into or out of the specified element by p_1 passing the limit p_1'. It will be convenient, and it is evidently allowable, to suppose dt so small that the quantities $\dot{p}_1\,dt$, $\dot{q}_1\,dt$, etc., which represent the increments of p_1, q_1, etc., in the time dt shall be infinitely small in comparison with the infinitesimal differences $p_1'' - p_1'$, $q_1'' - q_1'$, etc., which determine the magnitude of the element of extension-in-phase. The systems for which p_1 passes the limit p_1' in the interval dt are those for which at the commencement of this interval the value of p_1 lies between p_1' and $p_1' - \dot{p}_1\,dt$, as is evident if we consider separately the cases in which $\dot{p}_1$ is positive and negative. Those systems for which p_1 lies between these limits, and the other p's and q's between the limits specified in (9), will therefore pass into or out of the element considered according as $\dot{p}$ is positive or negative, unless indeed they also pass some other limit specified in (9) during the same interval of time. But the number which pass any two of these limits will be represented by an expression containing the square of dt as a factor, and is evidently negligible, when dt is sufficiently small, compared with the number which we are seeking to evaluate, and which (with neglect of terms containing dt^2) may be found by substituting $\dot{p}_1\,dt$ for $p_1'' - p_1'$ in (10) or for dp_1 in (11).

The expression

$$D\dot{p}_1\,dt\,dp_2 \ldots dp_n\,dq_1 \ldots dq_n \tag{13}$$

will therefore represent, according as it is positive or negative, the increase or decrease of the number of systems within the given limits which is due to systems passing the limit p_1'. A similar expression, in which however D and $\dot{p}$ will have slightly different values (being determined for p_1'' instead of p_1'), will represent the decrease or increase of the number of systems due to the passing of the limit p_1''. The difference of the two expressions, or

$$\frac{d(D\dot{p}_1)}{dp_1}\,dp_1 \ldots dp_n\,dq_1 \ldots dq_n\,dt \tag{14}$$

will represent algebraically the decrease of the number of systems within the limits due to systems passing the limits p_1' and p_1''.

The decrease in the number of systems within the limits due to systems passing the limits q_1' and q_1'' may be found in the same way. This will give

$$\left(\frac{d(D\dot{p}_1)}{dp_1} + \frac{d(D\dot{q}_1)}{dq_1}\right) dp_1 \ldots dp_n\, dq_1 \ldots dq_n\, dt \tag{15}$$

for the decrease due to passing the four limits p_1', p_1'', q_1', q_1''. But since the equations of motion (3) give

$$\frac{d\dot{p}_1}{dp_1} + \frac{d\dot{q}_1}{dq_1} = 0, \tag{16}$$

the expression reduces to

$$\left(\frac{dD}{dp_1}\dot{p}_1 + \frac{dD}{dq_1}\dot{q}_1\right) dp_1 \ldots dp_n\, dq_1 \ldots dq_n\, dt. \tag{17}$$

If we prefix Σ to denote summation relative to the suffixes $1 \ldots n$, we get the total decrease in the number of systems within the limits in the time dt. That is,

$$\Sigma \left(\frac{dD}{dp_1}\dot{p}_1 + \frac{dD}{dq_1}\dot{q}_1\right) dp_1 \ldots dp_n\, dq_1 \ldots dq_n\, dt = \\ - dD\, dp_1 \ldots dp_n\, dq_1 \ldots dq_n, \tag{18}$$

or

$$\left(\frac{dD}{dt}\right)_{p,q} = - \Sigma \left(\frac{dD}{dp_1}\dot{p}_1 + \frac{dD}{dq_1}\dot{q}_1\right), \tag{19}$$

where the suffix applied to the differential coefficient indicates that the p's and q's are to be regarded as constant in the differentiation. The condition of statistical equilibrium is therefore

$$\Sigma \left(\frac{dD}{dp_1}\dot{p}_1 + \frac{dD}{dq_1}\dot{q}_1\right) = 0. \tag{20}$$

If at any instant this condition is fulfilled for all values of the p's and q's, $(dD/dt)_{p,q}$ vanishes, and therefore the condition will continue to hold, and the distribution in phase will be permanent, so long as the external coördinates remain constant. But the statistical equilibrium would in general be disturbed by a change in the values of the external coördinates, which would alter the values of the $\dot{p}$'s as determined by equations (3), and thus disturb the relation expressed in the last equation.

If we write equation (19) in the form

$$\left(\frac{dD}{dt}\right)_{p,q} dt + \Sigma \left(\frac{dD}{dp_1}\dot{p}_1\, dt + \frac{dD}{dq_1}\dot{q}_1\, dt\right) = 0, \tag{21}$$

it will be seen to express a theorem of remarkable simplicity. Since D is a function of t, $p_1, \ldots p_n, q_1, \ldots q_n$, its complete differential will consist of parts due to the variations of all these quantities. Now the first term of the equation represents the increment of D due to an increment of t (with constant values of the p's and q's), and the rest of the first member represents the increments of D due to increments of the p's and q's, expressed by $\dot{p}_1\, dt$, $\dot{q}_1\, dt$, etc. But these are precisely the increments which the p's and q's receive in the movement of a system in the time dt. The whole expression represents the total increment of D for the varying phase of a moving system. We have therefore the theorem:—

*In an ensemble of mechanical systems identical in nature and subject to forces determined by identical laws, but distributed in phase in any continuous manner, the density-in-phase is constant in time for the varying phases of a moving system; provided, that the forces of a system are functions of its coördinates, either alone or with the time.**

This may be called the principle of *conservation of density-in-phase.* It may also be written

$$\left(\frac{dD}{dt}\right)_{a, \ldots h} = 0, \tag{22}$$

where $a, \ldots h$ represent the arbitrary constants of the integral equations of motion, and are suffixed to the differential co-

* The condition that the forces $F_1, \ldots F_n$ are functions of $q_1, \ldots q_n$ and a_1, a_2, etc., which last are functions of the time, is analytically equivalent to the condition that $F_1, \ldots F_n$ are functions of $q_1, \ldots q_n$ and the time. Explicit mention of the external coördinates, a_1, a_2, etc., has been made in the preceding pages, because our purpose will require us hereafter to consider these coördinates and the connected forces, A_1, A_2, etc., which represent the action of the systems on external bodies.

efficient to indicate that they are to be regarded as constant in the differentiation.

We may give to this principle a slightly different expression. Let us call the value of the integral

$$\int \dots \int dp_1 \dots dp_n \, dq_1 \dots dq_n \tag{23}$$

taken within any limits the *extension-in-phase* within those limits.

When the phases bounding an extension-in-phase vary in the course of time according to the dynamical laws of a system subject to forces which are functions of the coördinates either alone or with the time, the value of the extension-in-phase thus bounded remains constant. In this form the principle may be called the principle of *conservation of extension-in-phase.* In some respects this may be regarded as the most simple statement of the principle, since it contains no explicit reference to an ensemble of systems.

Since any extension-in-phase may be divided into infinitesimal portions, it is only necessary to prove the principle for an infinitely small extension. The number of systems of an ensemble which fall within the extension will be represented by the integral

$$\int \dots \int D \, dp_1 \dots dp_n \, dq_1 \dots dq_n.$$

If the extension is infinitely small, we may regard D as constant in the extension and write

$$D \int \dots \int dp_1 \dots dp_n \, dq_1 \dots dq_n$$

for the number of systems. The value of this expression must be constant in time, since no systems are supposed to be created or destroyed, and none can pass the limits, because the motion of the limits is identical with that of the systems. But we have seen that D is constant in time, and therefore the integral

$$\int \dots \int dp_1 \dots dp_n \, dq_1 \dots dq_n,$$

which we have called the extension-in-phase, is also constant in time.*

Since the system of coördinates employed in the foregoing discussion is entirely arbitrary, the values of the coördinates relating to any configuration and its immediate vicinity do not impose any restriction upon the values relating to other configurations. The fact that the quantity which we have called density-in-phase is constant in time for any given system, implies therefore that its value is independent of the coördinates which are used in its evaluation. For let the density-in-phase as evaluated for the same time and phase by one system of coördinates be D_1', and by another system D_2'. A system which at that time has that phase will at another time have another phase. Let the density as calculated for this second time and phase by a third system of coördinates be D_3''. Now we may imagine a system of coördinates which at and near the first configuration will coincide with the first system of coördinates, and at and near the second configuration will coincide with the third system of coördinates. This will give $D_1' = D_3''$. Again we may imagine a system of coördinates which at and near the first configuration will coincide with the second system of coördinates, and at and near the

* If we regard a phase as represented by a point in space of $2n$ dimensions, the changes which take place in the course of time in our ensemble of systems will be represented by a current in such space. This current will be steady so long as the external coördinates are not varied. In any case the current will satisfy a law which in its various expressions is analogous to the hydrodynamic law which may be expressed by the phrases *conservation of volumes* or *conservation of density about a moving point*, or by the equation

$$\frac{d\dot{x}}{dx} + \frac{d\dot{y}}{dy} + \frac{d\dot{z}}{dz} = 0.$$

The analogue in statistical mechanics of this equation, viz.,

$$\frac{d\dot{p}_1}{dp_1} + \frac{d\dot{q}_1}{dq_1} + \frac{d\dot{p}_2}{dp_2} + \frac{d\dot{q}_2}{dq_2} + \text{etc.} = 0,$$

may be derived directly from equations (3) or (6), and may suggest such theorems as have been enunciated, if indeed it is not regarded as making them intuitively evident. The somewhat lengthy demonstrations given above will at least serve to give precision to the notions involved, and familiarity with their use.

second configuration will coincide with the third system of coördinates. This will give $D_2' = D_3''$. We have therefore $D_1' = D_2'$.

It follows, or it may be proved in the same way, that the value of an extension-in-phase is independent of the system of coördinates which is used in its evaluation. This may easily be verified directly. If $q_1, \ldots q_n$, $Q_1, \ldots Q_n$ are two systems of coördinates, and $p_1, \ldots p_n$, $P_1, \ldots P_n$ the corresponding momenta, we have to prove that

$$\int \cdots \int dp_1 \ldots dp_n dq_1 \ldots dq_n = \int \cdots \int dP_1 \ldots dP_n dQ_1 \ldots dQ_n, \quad (24)$$

when the multiple integrals are taken within limits consisting of the same phases. And this will be evident from the principle on which we change the variables in a multiple integral, if we prove that

$$\frac{d(P_1, \ldots P_n, Q_1, \ldots Q_n)}{d(p_1, \ldots p_n, q_1, \ldots q_n)} = 1, \quad (25)$$

where the first member of the equation represents a Jacobian or functional determinant. Since all its elements of the form dQ/dp are equal to zero, the determinant reduces to a product of two, and we have to prove that

$$\frac{d(P_1, \ldots P_n)}{d(p_1, \ldots p_n)} \frac{d(Q_1, \ldots Q_n)}{d(q_1, \ldots q_n)} = 1. \quad (26)$$

We may transform any element of the first of these determinants as follows. By equations (2) and (3), and in view of the fact that the $\dot{Q}$'s are linear functions of the $\dot{q}$'s and therefore of the p's, with coefficients involving the q's, so that a differential coefficient of the form $d\dot{Q}_r/dp_y$ is function of the q's alone, we get*

* The form of the equation

$$\frac{d}{dp_y} \frac{d\epsilon_p}{d\dot{Q}_x} = \frac{d}{d\dot{Q}_x} \frac{d\epsilon_p}{dp_y}$$

in (27) reminds us of the fundamental identity in the differential calculus relating to the order of differentiation with respect to independent variables. But it will be observed that here the variables $\dot{Q}_x$ and p_y are *not* independent and that the proof depends on the *linear* relation between the $\dot{Q}$'s and the p's.

$$\frac{dP_x}{dp_y} = \frac{d}{dp_y}\frac{d\epsilon_p}{d\dot{Q}_x} = \sum_{r=1}^{r=n}\left(\frac{d^2\epsilon_p}{d\dot{Q}_r\, d\dot{Q}_x}\frac{d\dot{Q}_r}{dp_y}\right) =$$

$$\frac{d}{d\dot{Q}_x}\sum_{r=1}^{r=n}\left(\frac{d\epsilon_p}{d\dot{Q}_r}\frac{d\dot{Q}_r}{dp_y}\right) = \frac{d}{d\dot{Q}_x}\frac{d\epsilon_p}{dp_y} = \frac{d\dot{q}_y}{d\dot{Q}_x}. \qquad (27)$$

But since
$$\dot{q}_y = \sum_{r=1}^{r=n}\left(\frac{dq_y}{dQ_r}\dot{Q}_r\right),$$

$$\frac{d\dot{q}_y}{d\dot{Q}_x} = \frac{dq_y}{dQ_x}. \qquad (28)$$

Therefore,

$$\frac{d(P_1, \ldots P_n)}{d(p_1, \ldots p_n)} = \frac{d(\dot{q}_1, \ldots \dot{q}_n)}{d(\dot{Q}_1, \ldots \dot{Q}_n)} = \frac{d(q_1, \ldots q_n)}{d(Q_1, \ldots Q_n)}. \qquad (29)$$

The equation to be proved is thus reduced to

$$\frac{d(q_1, \ldots q_n)}{d(Q_1, \ldots Q_n)}\frac{d(Q_1, \ldots Q_n)}{d(q_1, \ldots q_n)} = 1, \qquad (30)$$

which is easily proved by the ordinary rule for the multiplication of determinants.

The numerical value of an extension-in-phase will however depend on the units in which we measure energy and time. For a product of the form $dp\, dq$ has the dimensions of energy multiplied by time, as appears from equation (2), by which the momenta are defined. Hence an extension-in-phase has the dimensions of the nth power of the product of energy and time. In other words, it has the dimensions of the nth power of *action*, as the term is used in the 'principle of *Least Action*.'

If we distinguish by accents the values of the momenta and coördinates which belong to a time t', the unaccented letters relating to the time t, the principle of the conservation of extension-in-phase may be written

$$\int \ldots \int dp_1 \ldots dp_n dq_1 \ldots dq_n = \int \ldots \int dp_1' \ldots dp_n' dq_1' \ldots dq_n', \qquad (31)$$

or more briefly

$$\int \ldots \int dp_1 \ldots dq_n = \int \ldots \int dp_1' \ldots dq_n', \qquad (32)$$

the limiting phases being those which belong to the same systems at the times t and t' respectively. But we have identically

$$\int \dots \int dp_1 \dots dq_n = \int \dots \int \frac{d(p_1 \dots q_n)}{d(p_1' \dots q_n')} dp_1' \dots dq_n'$$

for such limits. The principle of conservation of extension-in-phase may therefore be expressed in the form

$$\frac{d(p_1, \dots q_n)}{d(p_1' \dots q_n')} = 1. \tag{33}$$

This equation is easily proved directly. For we have identically

$$\frac{d(p_1, \dots q_n)}{d(p_1', \dots q_n)} = \frac{d(p_1, \dots q_n)}{d(p_1'', \dots q_n'')} \frac{d(p_1'', \dots q_n'')}{d(p_1', \dots q_n')},$$

where the double accents distinguish the values of the momenta and coördinates for a time t''. If we vary t, while t' and t'' remain constant, we have

$$\frac{d}{dt} \frac{d(p_1, \dots q_n)}{d(p_1', \dots q_n')} = \frac{d(p_1'', \dots q_n'')}{d(p_1', \dots q_n')} \frac{d}{dt} \frac{d(p_1, \dots q_n)}{d(p_1'', \dots q_n'')}. \tag{34}$$

Now since the time t'' is entirely arbitrary, nothing prevents us from making t'' identical with t at the moment considered. Then the determinant

$$\frac{d(p_1, \dots q_n)}{d(p_1'', \dots q_n'')}$$

will have unity for each of the elements on the principal diagonal, and zero for all the other elements. Since every term of the determinant except the product of the elements on the principal diagonal will have two zero factors, the differential of the determinant will reduce to that of the product of these elements, *i. e.*, to the sum of the differentials of these elements. This gives the equation

$$\frac{d}{dt} \frac{d(p_1, \dots q_n)}{d(p_1'', \dots q_n'')} = \frac{d\dot{p}_1}{dp_1''} \dots + \frac{d\dot{p}_n}{dp_n''} + \frac{d\dot{q}_1}{dq_1''} \dots + \frac{d\dot{q}_n}{dq_n''}.$$

Now since $t = t''$, the double accents in the second member of this equation may evidently be neglected. This will give, in virtue of such relations as (16),

$$\frac{d}{dt}\frac{d(p_1, \ldots q_n)}{d(p_1'', \ldots q_n'')} = 0,$$

which substituted in (34) will give

$$\frac{d}{dt}\frac{d(p_1, \ldots q_n)}{d(p_1', \ldots q_n')} = 0.$$

The determinant in this equation is therefore a constant, the value of which may be determined at the instant when $t = t'$, when it is evidently unity. Equation (33) is therefore demonstrated.

Again, if we write $a, \ldots h$ for a system of $2n$ arbitrary constants of the integral equations of motion, p_1, q_1, etc. will be functions of $a, \ldots h$, and t, and we may express an extension-in-phase in the form

$$\int \cdots \int \frac{d(p_1, \ldots q_n)}{d(a, \ldots h)}\, da \ldots dh. \tag{35}$$

If we suppose the limits specified by values of $a, \ldots h$, a system initially at the limits will remain at the limits. The principle of conservation of extension-in-phase requires that an extension thus bounded shall have a constant value. This requires that the determinant under the integral sign shall be constant, which may be written

$$\frac{d}{dt}\frac{d(p_1, \ldots q_n)}{d(a, \ldots h)} = 0. \tag{36}$$

This equation, which may be regarded as expressing the principle of conservation of extension-in-phase, may be derived directly from the identity

$$\frac{d(p_1, \ldots q_n)}{d(a, \ldots h)} = \frac{d(p_1, \ldots q_n)}{d(p_1', \ldots q_n')}\,\frac{d(p_1', \ldots q_n')}{d(a, \ldots h)}$$

in connection with equation (33).

Since the coördinates and momenta are functions of $a, \ldots h$, and t, the determinant in (36) must be a function of the same variables, and since it does not vary with the time, it must be a function of $a, \ldots h$ alone. We have therefore

$$\frac{d(p_1, \ldots q_n)}{d(a, \ldots h)} = \text{func.}\,(a, \ldots h). \tag{37}$$

It is the relative numbers of systems which fall within different limits, rather than the absolute numbers, with which we are most concerned. It is indeed only with regard to relative numbers that such discussions as the preceding will apply with literal precision, since the nature of our reasoning implies that the number of systems in the smallest element of space which we consider is very great. This is evidently inconsistent with a finite value of the total number of systems, or of the density-in-phase. Now if the value of D is infinite, we cannot speak of any definite number of systems within any finite limits, since all such numbers are infinite. But the ratios of these infinite numbers may be perfectly definite. If we write N for the total number of systems, and set

$$P = \frac{D}{N}, \tag{38}$$

P may remain finite, when N and D become infinite. The integral

$$\int \ldots \int P \, dp_1 \ldots dq_n \tag{39}$$

taken within any given limits, will evidently express the ratio of the number of systems falling within those limits to the whole number of systems. This is the same thing as the *probability* that an unspecified system of the ensemble (*i. e.* one of which we only know that it belongs to the ensemble) will lie within the given limits. The product

$$P \, dp_1 \ldots dq_n \tag{40}$$

expresses the probability that an unspecified system of the ensemble will be found in the element of extension-in-phase $dp_1 \ldots dq_n$. We shall call P the *coefficient of probability* of the phase considered. Its natural logarithm we shall call the *index of probability* of the phase, and denote it by the letter η.

If we substitute NP and Ne^η for D in equation (19), we get

$$\left(\frac{dP}{dt}\right)_{p,q} = - \Sigma \left(\frac{dP}{dp_1} \dot{p}_1 + \frac{dP}{dq_1} \dot{dq_1}\right), \tag{41}$$

and

$$\left(\frac{d\eta}{dt}\right)_{p,q} = - \Sigma \left(\frac{d\eta}{dp_1} \dot{p}_1 + \frac{d\eta}{dq_1} \dot{dq_1}\right). \tag{42}$$

The condition of statistical equilibrium may be expressed by equating to zero the second member of either of these equations.

The same substitutions in (22) give

$$\left(\frac{dP}{dt}\right)_{a,\ldots h} = 0, \tag{43}$$

and

$$\left(\frac{d\eta}{dt}\right)_{a,\ldots h} = 0. \tag{44}$$

That is, the values of P and η, like those of D, are constant in time for moving systems of the ensemble. From this point of view, the principle which otherwise regarded has been called the principle of conservation of density-in-phase or conservation of extension-in-phase, may be called the principle of conservation of the coefficient (or index) of probability of a phase varying according to dynamical laws, or more briefly, the principle of *conservation of probability of phase.* It is subject to the limitation that the forces must be functions of the coördinates of the system either alone or with the time.

The application of this principle is not limited to cases in which there is a formal and explicit reference to an ensemble of systems. Yet the conception of such an ensemble may serve to give precision to notions of probability. It is in fact customary in the discussion of probabilities to describe anything which is imperfectly known as something taken at random from a great number of things which are completely described. But if we prefer to avoid any reference to an ensemble of systems, we may observe that the probability that the phase of a system falls within certain limits at a certain time, is equal to the probability that at some other time the phase will fall within the limits formed by phases corresponding to the first. For either occurrence necessitates the other. That is, if we write P' for the coefficient of probability of the phase $p_1', \ldots q_n'$ at the time t', and P'' for that of the phase $p_1'', \ldots q_n''$ at the time t'',

$$\int \ldots \int P' \, dq_1' \ldots dq_n' = \int \ldots \int P'' \, dp_1'' \ldots dq_n'', \qquad (45)$$

where the limits in the two cases are formed by corresponding phases. When the integrations cover infinitely small variations of the momenta and coördinates, we may regard P' and P'' as constant in the integrations and write

$$P' \int \ldots \int dp_1' \ldots dq_n'' = P'' \int \ldots \int dp_1'' \ldots dq_n''.$$

Now the principle of the conservation of extension-in-phase, which has been proved (viz., in the second demonstration given above) independently of any reference to an ensemble of systems, requires that the values of the multiple integrals in this equation shall be equal. This gives

$$P'' = P'.$$

With reference to an important class of cases this principle may be enunciated as follows.

When the differential equations of motion are exactly known, but the constants of the integral equations imperfectly determined, the coefficient of probability of any phase at any time is equal to the coefficient of probability of the corresponding phase at any other time. By corresponding phases are meant those which are calculated for different times from the same values of the arbitrary constants of the integral equations.

Since the sum of the probabilities of all possible cases is necessarily unity, it is evident that we must have

$$\int_{\text{phases}}^{\text{all}} \ldots \int P \, dp_1 \ldots dq_n = 1, \qquad (46)$$

where the integration extends over all phases. This is indeed only a different form of the equation

$$N = \int_{\text{phases}}^{\text{all}} \ldots \int D \, dp_1 \ldots dq_n,$$

which we may regard as defining N.

The values of the coefficient and index of probability of phase, like that of the density-in-phase, are independent of the system of coördinates which is employed to express the distribution in phase of a given ensemble.

In dimensions, the coefficient of probability is the reciprocal of an extension-in-phase, that is, the reciprocal of the nth power of the product of time and energy. The index of probability is therefore affected by an additive constant when we change our units of time and energy. If the unit of time is multiplied by c_t and the unit of energy is multiplied by c_ϵ, all indices of probability relating to systems of n degrees of freedom will be increased by the addition of

$$n \log c_t + n \log c_\epsilon. \tag{47}$$

CHAPTER II.

APPLICATION OF THE PRINCIPLE OF CONSERVATION OF EXTENSION-IN-PHASE TO THE THEORY OF ERRORS.

LET us now proceed to combine the principle which has been demonstrated in the preceding chapter and which in its different applications and regarded from different points of view has been variously designated as the conservation of density-in-phase, or of extension-in-phase, or of probability of phase, with those approximate relations which are generally used in the 'theory of errors.'

We suppose that the differential equations of the motion of a system are exactly known, but that the constants of the integral equations are only approximately determined. It is evident that the probability that the momenta and coördinates at the time t' fall between the limits p_1' and $p_1' + dp_1'$, q_1' and $q_1' + dq_1'$, etc., may be expressed by the formula

$$e^{\eta'}\, dp_1' \ldots dq_n', \tag{48}$$

where η' (the index of probability for the phase in question) is a function of the coördinates and momenta and of the time.

Let Q_1', P_1', etc. be the values of the coördinates and momenta which give the maximum value to η', and let the general value of η' be developed by Taylor's theorem according to ascending powers and products of the differences $p_1' - P_1'$, $q_1' - Q_1'$, etc., and let us suppose that we have a sufficient approximation without going beyond terms of the second degree in these differences. We may therefore set

$$\eta' = c - F', \tag{49}$$

where c is independent of the differences $p_1' - P_1'$, $q_1' - Q_1'$, etc., and F' is a homogeneous quadratic function of these differences. The terms of the first degree vanish in virtue of the maximum condition, which also requires that F' must have a positive value except when all the differences mentioned vanish. If we set

$$C = e^c, \tag{50}$$

we may write for the probability that the phase lies within the limits considered

$$Ce^{-F'}\, dp_1' \ldots dq_n'. \tag{51}$$

C is evidently the maximum value of the coefficient of probability at the time considered.

In regard to the degree of approximation represented by these formulæ, it is to be observed that we suppose, as is usual in the 'theory of errors,' that the determination (explicit or implicit) of the constants of motion is of such precision that the coefficient of probability $e^{\eta'}$ or $Ce^{-F'}$ is practically zero except for very small values of the differences $p_1' - P_1'$, $q_1' - Q_1'$, etc. For very small values of these differences the approximation is evidently in general sufficient, for larger values of these differences the value of $Ce^{-F'}$ will be sensibly zero, as it should be, and in this sense the formula will represent the facts.

We shall suppose that the forces to which the system is subject are functions of the coördinates either alone or with the time. The principle of conservation of probability of phase will therefore apply, which requires that at any other time (t'') the maximum value of the coefficient of probability shall be the same as at the time t', and that the phase (P_1'', Q_1'', etc.) which has this greatest probability-coefficient, shall be that which corresponds to the phase (P_1', Q_1', etc.), *i. e.*, which is calculated from the same values of the constants of the integral equations of motion.

We may therefore write for the probability that the phase at the time t'' falls within the limits p_1'' and $p_1'' + dp_1''$, q_1'' and $q_1'' + dq_1''$, etc.,

$$Ce^{-F''}\, dp_1'' \ldots dq_n'', \tag{52}$$

where C represents the same value as in the preceding formula, viz., the constant value of the maximum coefficient of probability, and F'' is a quadratic function of the differences $p_1'' - P_1''$, $q_1'' - Q_1''$, etc., the phase (P_1'', Q_1'' etc.) being that which at the time t'' corresponds to the phase (P_1', Q_1', etc.) at the time t'.

Now we have necessarily

$$\int \ldots \int C e^{-F'} dp_1' \ldots dq_n' = \int \ldots \int C e^{-F''} dp_1'' \ldots dq_n'' = 1, \quad (53)$$

when the integration is extended over all possible phases. It will be allowable to set $\pm \infty$ for the limits of all the coördinates and momenta, not because these values represent the actual limits of possible phases, but because the portions of the integrals lying outside of the limits of all possible phases will have sensibly the value zero. With $\pm \infty$ for limits, the equation gives

$$\frac{C\pi^n}{\sqrt{f'}} = \frac{C\pi^n}{\sqrt{f''}} = 1, \quad (54)$$

where f' is the discriminant* of F', and f'' that of F''. This discriminant is therefore constant in time, and like C an absolute invariant in respect to the system of coördinates which may be employed. In dimensions, like C^2, it is the reciprocal of the 2nth power of the product of energy and time.

Let us see precisely how the functions F' and F'' are related. The principle of the conservation of the probability-coefficient requires that any values of the coördinates and momenta at the time t' shall give the function F' the same value as the corresponding coördinates and momenta at the time t'' give to F''. Therefore F'' may be derived from F' by substituting for $p_1', \ldots q_n'$ their values in terms of $p_1'', \ldots q_1''$. Now we have approximately

* This term is used to denote the determinant having for elements on the principal diagonal the coefficients of the squares in the quadratic function F', and for its other elements the halves of the coefficients of the products in F'.

$$\left.\begin{array}{l} p_1' - P_1' = \dfrac{dP_1'}{dP_1''}(p_1'' - P_1'') \ldots + \dfrac{dP_1'}{dQ_n''}(q_n'' - Q_n'') \\ \ldots\ldots\ldots\ldots\ldots\ldots\ldots\ldots\ldots \\ q_n' - Q_n' = \dfrac{dQ_n'}{dP_1''}(p_1'' - P_1'') \ldots + \dfrac{dQ_n'}{dQ_n''}(q_n'' - Q_n''), \end{array}\right\} \qquad (55)$$

and as in F'' terms of higher degree than the second are to be neglected, these equations may be considered accurate for the purpose of the transformation required. Since by equation (33) the eliminant of these equations has the value unity, the discriminant of F'' will be equal to that of F', as has already appeared from the consideration of the principle of conservation of probability of phase, which is, in fact, essentially the same as that expressed by equation (33).

At the time t', the phases satisfying the equation

$$F' = k, \qquad (56)$$

where k is any positive constant, have the probability-coefficient $C e^{-k}$. At the time t'', the corresponding phases satisfy the equation

$$F'' = k, \qquad (57)$$

and have the same probability-coefficient. So also the phases within the limits given by one or the other of these equations are corresponding phases, and have probability-coefficients greater than $C e^{-k}$, while phases without these limits have less probability-coefficients. The probability that the phase at the time t' falls within the limits $F' = k$ is the same as the probability that it falls within the limits $F'' = k$ at the time t'', since either event necessitates the other. This probability may be evaluated as follows. We may omit the accents, as we need only consider a single time. Let us denote the extension-in-phase within the limits $F = k$ by U, and the probability that the phase falls within these limits by R, also the extension-in-phase within the limits $F = 1$ by U_1. We have then by definition

$$U = \int \ldots \int^{F=k} dp_1 \ldots dq_n, \qquad (58)$$

$$R = \int \overset{F=k}{\ldots} \int C\, e^{-F}\, dp_1 \ldots dq_n, \tag{59}$$

$$U_1 = \int \overset{F=1}{\ldots} \int dp_1 \ldots dq_n. \tag{60}$$

But since F is a homogeneous quadratic function of the differences

$$p_1 - P_1, p_2 - P_2, \ldots q_n - Q_n,$$

we have identically

$$\int \overset{F=k}{\ldots} \int d(p_1 - P_1) \ldots d(q_n - Q_n)$$

$$= \int \overset{kF=k}{\ldots} \int k^n\, d(p_1 - P_1) \ldots d(q_n - Q_n)$$

$$= k^n \int \overset{F=1}{\ldots} \int d(p_1 - P_1) \ldots d(q_n - Q_n).$$

That is $$U = k^n\, U_1, \tag{61}$$

whence $$dU = U_1\, n\, k^{n-1}\, dk. \tag{62}$$

But if k varies, equations (58) and (59) give

$$dU = \int \overset{F=k+dk}{\underset{F=k}{\ldots}} \int dp_1 \ldots dq_n \tag{63}$$

$$dR = \int \overset{F=k+dk}{\underset{F=k}{\ldots}} \int C\, e^{-F}\, dp_1 \ldots dq_n \tag{64}$$

Since the factor $C e^{-F}$ has the constant value $C e^{-k}$ in the last multiple integral, we have

$$dR = C\, e^{-k}\, dU = C\, U_1\, n\, e^{-k}\, k^{n-1}\, dk, \tag{65}$$

$$R = - C\, U_1 \lfloor\!\underline{n}\, e^{-k} \left(1 + k + \frac{k^2}{2} + \ldots + \frac{k^{n-1}}{\lfloor\!\underline{n-1}}\right) + \text{const.} \tag{66}$$

We may determine the constant of integration by the condition that R vanishes with k. This gives

$$R = C\,U_1 \lfloor n - C\,U_1 \lfloor n\, e^{-k}\left(1 + k + \frac{k^2}{2} + \ldots + \frac{k^{n-1}}{\lfloor n-1}\right). \tag{67}$$

We may determine the value of the constant U_1 by the condition that $R = 1$ for $k = \infty$. This gives $C\,U_1 \lfloor n = 1$, and

$$R = 1 - e^{-k}\left(1 + k + \frac{k^2}{2} \ldots + \frac{k^{n-1}}{\lfloor n-1}\right), \tag{68}$$

$$U = \frac{k^n}{C \lfloor n}. \tag{69}$$

It is worthy of notice that the form of these equations depends only on the number of degrees of freedom of the system, being in other respects independent of its dynamical nature, except that the forces must be functions of the coördinates either alone or with the time.

If we write

$$k_{R=\frac{1}{2}}$$

for the value of k which substituted in equation (68) will give $R = \frac{1}{2}$, the phases determined by the equation

$$F = k_{R=\frac{1}{2}} \tag{70}$$

will have the following properties.

The probability that the phase falls within the limits formed by these phases is greater than the probability that it falls within any other limits enclosing an equal extension-in-phase. It is equal to the probability that the phase falls without the same limits.

These properties are analogous to those which in the theory of errors in the determination of a single quantity belong to values expressed by $A \pm a$, when A is the most probable value, and a the 'probable error.'

CHAPTER III.

APPLICATION OF THE PRINCIPLE OF CONSERVATION OF EXTENSION-IN-PHASE TO THE INTEGRATION OF THE DIFFERENTIAL EQUATIONS OF MOTION.*

WE have seen that the principle of conservation of extension-in-phase may be expressed as a differential relation between the coördinates and momenta and the arbitrary constants of the integral equations of motion. Now the integration of the differential equations of motion consists in the determination of these constants as functions of the coördinates and momenta with the time, and the relation afforded by the principle of conservation of extension-in-phase may assist us in this determination.

It will be convenient to have a notation which shall not distinguish between the coördinates and momenta. If we write $r_1 \ldots r_{2n}$ for the coördinates and momenta, and $a \ldots h$ as before for the arbitrary constants, the principle of which we wish to avail ourselves, and which is expressed by equation (37), may be written

$$\frac{d(r_1, \ldots r_{2n})}{d(a, \ldots h)} = \text{func.}(a, \ldots h). \tag{71}$$

Let us first consider the case in which the forces are determined by the coördinates alone. Whether the forces are 'conservative' or not is immaterial. Since the differential equations of motion do not contain the time (t) in the finite form, if we eliminate dt from these equations, we obtain $2n-1$ equations in $r_1, \ldots r_{2n}$ and their differentials, the integration of which will introduce $2n-1$ arbitrary constants which we shall call $b \ldots h$. If we can effect these integrations, the

* See Boltzmann: "Zusammenhang zwischen den Sätzen über das Verhalten mehratomiger Gasmolecüle mit Jacobi's Princip des letzten Multiplicators. Sitzb. der Wiener Akad., Bd. LXIII, Abth. II., S. 679, (1871).

remaining constant (a) will then be introduced in the final integration, (viz., that of an equation containing dt,) and will be added to or subtracted from t in the integral equation. Let us have it subtracted from t. It is evident then that

$$\frac{dr_1}{da} = -\dot{r}_1, \quad \frac{dr_2}{da} = -\dot{r}_2, \quad \text{etc.} \tag{72}$$

Moreover, since $b, \ldots h$ and $t - a$ are independent functions of $r_1, \ldots r_{2n}$, the latter variables are functions of the former. The Jacobian in (71) is therefore function of $b, \ldots h$, and $t - a$, and since it does not vary with t it cannot vary with a. We have therefore in the case considered, viz., where the forces are functions of the coördinates alone,

$$\frac{d(r_1, \ldots r_{2n})}{d(a, \ldots h)} = \text{func.}\,(b, \ldots h). \tag{73}$$

Now let us suppose that of the first $2n - 1$ integrations we have accomplished all but one, determining $2n - 2$ arbitrary constants (say $c, \ldots h$) as functions of $r_1, \ldots r_{2n}$, leaving b as well as a to be determined. Our $2n - 2$ finite equations enable us to regard all the variables $r_1, \ldots r_{2n}$, and all functions of these variables as functions of two of them, (say r_1 and r_2,) with the arbitrary constants $c, \ldots h$. To determine b, we have the following equations for constant values of $c, \ldots h$.

$$dr_1 = \frac{dr_1}{da}\,da + \frac{dr_1}{db}\,db,$$

$$dr_2 = \frac{dr_2}{da}\,da + \frac{dr_2}{db}\,db,$$

whence

$$\frac{d(r_1, r_2)}{d(a, b)}\,db = -\frac{dr_2}{da}\,dr_1 + \frac{dr_1}{da}\,dr_2. \tag{74}$$

Now, by the ordinary formula for the change of variables,

$$\int \cdots \int \frac{d(r_1, r_2)}{d(a, b)}\,da\,db\,dr_3 \ldots dr_{2n} = \int \cdots \int dr_1 \ldots dr_{2n}$$

$$= \int \cdots \int \frac{d(r_1, \ldots r_{2n})}{d(a, \ldots h)}\,da \ldots dh$$

$$= \int \cdots \int \frac{d(r_1, \ldots r_{2n})}{d(a, \ldots h)} \frac{d(c, \ldots h)}{d(r_3, \ldots r_{2n})}\,da\,db\,dr_3 \ldots dr_{2n},$$

where the limits of the multiple integrals are formed by the same phases. Hence

$$\frac{d(r_1, r_2)}{d(a, b)} = \frac{d(r_1, \ldots r_{2n})}{d(a, \ldots h)} \frac{d(c, \ldots h)}{d(r_3, \ldots r_{2n})}, \tag{75}$$

With the aid of this equation, which is an identity, and (72), we may write equation (74) in the form

$$\frac{d(r_1, \ldots r_{2n})}{d(a, \ldots h)} \frac{d(c, \ldots h)}{d(r_3, \ldots r_{2n})} db = \dot{r}_2\, dr_1 - \dot{r}_1\, dr_2. \tag{76}$$

The separation of the variables is now easy. The differential equations of motion give $\dot{r}_1$ and $\dot{r}_2$ in terms of $r_1, \ldots r_{2n}$. The integral equations already obtained give $c, \ldots h$ and therefore the Jacobian $d(c, \ldots h)/d(r_3, \ldots r_{2n})$, in terms of the same variables. But in virtue of these same integral equations, we may regard functions of $r_1, \ldots r_{2n}$ as functions of r_1 and r_2 with the constants $c, \ldots h$. If therefore we write the equation in the form

$$\frac{d(r_1, \ldots r_{2n})}{d(a, \ldots h)} db = \frac{\dot{r}_2}{\dfrac{d(c, \ldots h)}{d(r_3, \ldots r_{2n})}} dr_1 - \frac{\dot{r}_1}{\dfrac{d(c, \ldots h)}{d(r_3, \ldots r_{2n})}} dr_2, \tag{77}$$

the coefficients of dr_1 and dr_2 may be regarded as known functions of r_1 and r_2 with the constants $c, \ldots h$. The coefficient of db is by (73) a function of $b, \ldots h$. It is not indeed a known function of these quantities, but since $c, \ldots h$ are regarded as constant in the equation, we know that the first member must represent the differential of some function of $b, \ldots h$, for which we may write b'. We have thus

$$db' = \frac{\dot{r}_2}{\dfrac{d(c, \ldots h)}{d(r_3, \ldots r_{2n})}} dr_1 - \frac{\dot{r}_1}{\dfrac{d(c_1, \ldots h)}{d(r_3, \ldots r_{2n})}} dr_2, \tag{78}$$

which may be integrated by quadratures and gives b' as functions of $r_1, r_2, \ldots c, \ldots h$, and thus as function of $r_1, \ldots r_{2n}$.

This integration gives us the last of the arbitrary constants which are functions of the coördinates and momenta without the time. The final integration, which introduces the remaining constant (a), is also a quadrature, since the equation to be integrated may be expressed in the form

$$dt = F(r_1)\, dr_1.$$

Now, apart from any such considerations as have been adduced, if we limit ourselves to the changes which take place in time, we have identically

$$\dot{r}_2\, dr_1 - \dot{r}_1\, dr_2 = 0,$$

and $\dot{r}_1$ and $\dot{r}_2$ are given in terms of $r_1, \ldots r_{2n}$ by the differential equations of motion. When we have obtained $2n - 2$ integral equations, we may regard $\dot{r}_2$ and $\dot{r}_1$ as known functions of r_1 and r_2. The only remaining difficulty is in integrating this equation. If the case is so simple as to present no difficulty, or if we have the skill or the good fortune to perceive that the multiplier

$$\frac{1}{\dfrac{d(c, \ldots h)}{d(r_3, \ldots r_{2n})}}, \tag{79}$$

or any other, will make the first member of the equation an exact differential, we have no need of the rather lengthy considerations which have been adduced. The utility of the principle of conservation of extension-in-phase is that it supplies a 'multiplier' which renders the equation integrable, and which it might be difficult or impossible to find otherwise.

It will be observed that the function represented by b' is a particular case of that represented by b. The system of arbitrary constants $a, b', c \ldots h$ has certain properties notable for simplicity. If we write b' for b in (77), and compare the result with (78), we get

$$\frac{d(r_1 \ldots r_{2n})}{d(a, b', c, \ldots h)} = 1. \tag{80}$$

Therefore the multiple integral

$$\int \ldots \int da\, db'\, dc \ldots dh \tag{81}$$

taken within limits formed by phases regarded as contemporaneous represents the extension-in-phase within those limits.

The case is somewhat different when the forces are not determined by the coördinates alone, but are functions of the coördinates with the time. All the arbitrary constants of the integral equations must then be regarded in the general case as functions of $r_1, \ldots r_{2n}$, and t. We cannot use the principle of conservation of extension-in-phase until we have made $2n-1$ integrations. Let us suppose that the constants $b, \ldots h$ have been determined by integration in terms of $r_1, \ldots r_{2n}$, and t, leaving a single constant (a) to be thus determined. Our $2n-1$ finite equations enable us to regard all the variables $r_1, \ldots r_{2n}$ as functions of a single one, say r_1.

For constant values of $b, \ldots h$, we have

$$dr_1 = \frac{dr_1}{da}\, da + \dot{r}_1\, dt. \tag{82}$$

Now

$$\int \cdots \int \frac{dr_1}{da}\, da\, dr_2 \ldots dr_{2n} = \int \cdots \int dr_1 \ldots dr_{2n}$$

$$= \int \cdots \int \frac{d(r_1, \ldots r_{2n})}{d(a, \ldots h)}\, da \ldots dh$$

$$= \int \cdots \int \frac{d(r_1, \ldots r_{2n})}{d(a, \ldots h)} \frac{d(b, \ldots h)}{d(r_2, \ldots r_{2n})}\, da\, dr_2 \ldots dr_{2n},$$

where the limits of the integrals are formed by the same phases. We have therefore

$$\frac{dr_1}{da} = \frac{d(r_1, \ldots r_{2n})}{d(a, \ldots h)} \frac{d(b, \ldots h)}{d(r_2, \ldots r_{2n})}, \tag{83}$$

by which equation (82) may be reduced to the form

$$\frac{d(r_1, \ldots r_{2n})}{d(a, \ldots h)}\, da = \frac{1}{\dfrac{d(b, \ldots h)}{d(r_2, \ldots r_{2n})}}\, dr_1 - \frac{\dot{r}_1}{\dfrac{d(b, \ldots h)}{d(r_2, \ldots r_{2n})}}\, dt. \tag{84}$$

Now we know by (71) that the coefficient of da is a function of $a, \ldots h$. Therefore, as $b, \ldots h$ are regarded as constant in the equation, the first number represents the differential of a function of $a, \dots h$, which we may denote by a'. We have then

$$da' = \frac{1}{\frac{d(b, \dots h)}{d(r_2, \dots r_{2n})}} dr_1 - \frac{\dot{r}_1'}{\frac{d(b, \dots h)}{d(r_2, \dots r_{2n})}} dt, \tag{85}$$

which may be integrated by quadratures. In this case we may say that the principle of conservation of extension-in-phase has supplied the 'multiplier'

$$\frac{1}{\frac{d(b, \dots h)}{d(r_2, \dots r_{2n})}} \tag{86}$$

for the integration of the equation

$$dr_1 - \dot{r}_1 \, dt = 0. \tag{87}$$

The system of arbitrary constants $a', b, \dots h$ has evidently the same properties which were noticed in regard to the system $a, b', \dots h$.

CHAPTER IV.

ON THE DISTRIBUTION IN PHASE CALLED CANONICAL, IN WHICH THE INDEX OF PROBABILITY IS A LINEAR FUNCTION OF THE ENERGY.

LET us now give our attention to the statistical equilibrium of ensembles of conservation systems, especially to those cases and properties which promise to throw light on the phenomena of thermodynamics.

The condition of statistical equilibrium may be expressed in the form *

$$\Sigma\left(\frac{dP}{dp_1}\dot{p}_1 + \frac{dP}{dq_1}\dot{q}_1\right) = 0, \tag{88}$$

where P is the coefficient of probability, or the quotient of the density-in-phase by the whole number of systems. To satisfy this condition, it is necessary and sufficient that P should be a function of the p's and q's (the momenta and coördinates) which does not vary with the time in a moving system. In all cases which we are now considering, the energy, or any function of the energy, is such a function.

$$P = \text{func.}\ (\epsilon)$$

will therefore satisfy the equation, as indeed appears identically if we write it in the form

$$\Sigma\left(\frac{dP}{dq_1}\frac{d\epsilon}{dp_1} - \frac{dP}{dp_1}\frac{d\epsilon}{dq_1}\right) = 0.$$

There are, however, other conditions to which P is subject, which are not so much conditions of statistical equilibrium, as conditions implicitly involved in the definition of the coeffi-

* See equations (20), (41), (42), also the paragraph following equation (20). The positions of any external bodies which can affect the systems are here supposed uniform for all the systems and constant in time.

cient of probability, whether the case is one of equilibrium or not. These are: that P should be single-valued, and neither negative nor imaginary for any phase, and that expressed by equation (46), viz.,

$$\int_{\text{phases}}^{\text{all}} \cdots \int P\, dp_1 \ldots dq_n = 1. \tag{89}$$

These considerations exclude

$$P = \epsilon \times \text{constant},$$

as well as

$$P = \text{constant},$$

as cases to be considered.

The distribution represented by

$$\eta = \log P = \frac{\psi - \epsilon}{\Theta}, \tag{90}$$

or

$$P = e^{\frac{\psi - \epsilon}{\Theta}}, \tag{91}$$

where Θ and ψ are constants, and Θ positive, seems to represent the most simple case conceivable, since it has the property that when the system consists of parts with separate energies, the laws of the distribution in phase of the separate parts are of the same nature,— a property which enormously simplifies the discussion, and is the foundation of extremely important relations to thermodynamics. The case is not rendered less simple by the divisor Θ, (a quantity of the same dimensions as ϵ,) but the reverse, since it makes the distribution independent of the units employed. The negative sign of ϵ is required by (89), which determines also the value of ψ for any given Θ, viz.,

$$e^{-\frac{\psi}{\Theta}} = \int_{\text{phases}}^{\text{all}} \cdots \int e^{-\frac{\epsilon}{\Theta}}\, dp_1 \ldots dq_n. \tag{92}$$

When an ensemble of systems is distributed in phase in the manner described, *i. e.*, when the index of probability is a linear function of the energy, we shall say that the ensemble is *canonically distributed*, and shall call the divisor of the energy (Θ) the *modulus* of distribution.

The fractional part of an ensemble canonically distributed which lies within any given limits of phase is therefore represented by the multiple integral

$$\int \ldots \int e^{\frac{\psi-\epsilon}{\Theta}} dp_1 \ldots dq_n \tag{93}$$

taken within those limits. We may express the same thing by saying that the multiple integral expresses the probability that an unspecified system of the ensemble (*i. e.*, one of which we only know that it belongs to the ensemble) falls within the given limits.

Since the value of a multiple integral of the form (23) (which we have called an extension-in-phase) bounded by any given phases is independent of the system of coördinates by which it is evaluated, the same must be true of the multiple integral in (92), as appears at once if we divide up this integral into parts so small that the exponential factor may be regarded as constant in each. The value of ψ is therefore independent of the system of coördinates employed.

It is evident that ψ might be defined as the energy for which the coefficient of probability of phase has the value unity. Since however this coefficient has the dimensions of the inverse nth power of the product of energy and time,* the energy represented by ψ is not independent of the units of energy and time. But when these units have been chosen, the definition of ψ will involve the same arbitrary constant as ϵ, so that, while in any given case the numerical values of ψ or ϵ will be entirely indefinite until the zero of energy has also been fixed for the system considered, the difference $\psi - \epsilon$ will represent a perfectly definite amount of energy, which is entirely independent of the zero of energy which we may choose to adopt.

* See Chapter I, p. 19.

It is evident that the canonical distribution is entirely determined by the modulus (considered as a quantity of energy) and the nature of the system considered, since when equation (92) is satisfied the value of the multiple integral (93) is independent of the units and of the coördinates employed, and of the zero chosen for the energy of the system.

In treating of the canonical distribution, we shall always suppose the multiple integral in equation (92) to have a finite value, as otherwise the coefficient of probability vanishes, and the law of distribution becomes illusory. This will exclude certain cases, but not such apparently, as will affect the value of our results with respect to their bearing on thermodynamics. It will exclude, for instance, cases in which the system or parts of it can be distributed in unlimited space (or in a space which has limits, but is still infinite in volume), while the energy remains beneath a finite limit. It also excludes many cases in which the energy can decrease without limit, as when the system contains material points which attract one another inversely as the squares of their distances. Cases of material points attracting each other inversely as the distances would be excluded for some values of Θ, and not for others. The investigation of such points is best left to the particular cases. For the purposes of a general discussion, it is sufficient to call attention to the assumption implicitly involved in the formula (92).*

The modulus Θ has properties analogous to those of temperature in thermodynamics. Let the system A be defined as one of an ensemble of systems of m degrees of freedom distributed in phase with a probability-coefficient

$$e^{\frac{\psi_A - \epsilon_A}{\Theta}},$$

* It will be observed that similar limitations exist in thermodynamics. In order that a mass of gas can be in thermodynamic equilibrium, it is necessary that it be enclosed. There is no thermodynamic equilibrium of a (finite) mass of gas in an infinite space. Again, that two attracting particles should be able to do an infinite amount of work in passing from one configuration (which is regarded as possible) to another, is a notion which, although perfectly intelligible in a mathematical formula, is quite foreign to our ordinary conceptions of matter.

and the system B as one of an ensemble of systems of n degrees of freedom distributed in phase with a probability-coefficient

$$e^{\frac{\psi_B-\epsilon_B}{\Theta}},$$

which has the same modulus. Let $q_1, \ldots q_m, p_1, \ldots p_m$ be the coördinates and momenta of A, and $q_{m+1}, \ldots q_{m+n}, p_{m+1}, \ldots p_{m+n}$ those of B. Now we may regard the systems A and B as together forming a system C, having $m+n$ degrees of freedom, and the coördinates and momenta $q_1, \ldots q_{m+n}, p_1, \ldots p_{m+n}$. The probability that the phase of the system C, as thus defined, will fall within the limits

$$dp_1, \ldots dp_{m+n}, dq_1, \ldots dq_{m+n}$$

is evidently the product of the probabilities that the systems A and B will each fall within the specified limits, viz.,

$$e^{\frac{\psi_A+\psi_B-\epsilon_A-\epsilon_B}{\Theta}} dp_1 \ldots dp_{m+n}\, dq_1 \ldots dq_{m+n}. \tag{94}$$

We may therefore regard C as an undetermined system of an ensemble distributed with the probability-coefficient

$$e^{\frac{\psi_A+\psi_B-(\epsilon_A+\epsilon_B)}{\Theta}}, \tag{95}$$

an ensemble which might be defined as formed by combining each system of the first ensemble with each of the second. But since $\epsilon_A + \epsilon_B$ is the energy of the whole system, and ψ_A and ψ_B are constants, the probability-coefficient is of the general form which we are considering, and the ensemble to which it relates is in statistical equilibrium and is canonically distributed.

This result, however, so far as statistical equilibrium is concerned, is rather nugatory, since conceiving of separate systems as forming a single system does not create any interaction between them, and if the systems combined belong to ensembles in statistical equilibrium, to say that the ensemble formed by such combinations as we have supposed is in statistical equilibrium, is only to repeat the data in different words. Let us therefore suppose that in forming the system C we add certain forces acting between A and B, and having the force-function $-\epsilon_{AB}$. The energy of the system C is now $\epsilon_A + \epsilon_B + \epsilon_{AB}$, and an ensemble of such systems distributed with a density proportional to

$$e^{\frac{-(\epsilon_A+\epsilon_B+\epsilon_{AB})}{\Theta}} \tag{96}$$

would be in statistical equilibrium. Comparing this with the probability-coefficient of C given above (95), we see that if we suppose ϵ_{AB} (or rather the variable part of this term when we consider all possible configurations of the systems A and B) to be infinitely small, the actual distribution in phase of C will differ infinitely little from one of statistical equilibrium, which is equivalent to saying that its distribution in phase will vary infinitely little even in a time indefinitely prolonged.* The case would be entirely different if A and B belonged to ensembles having different moduli, say Θ_A and Θ_B. The probability-coefficient of C would then be

$$e^{\frac{\psi_A-\epsilon_A}{\Theta_A}+\frac{\psi_B-\epsilon_B}{\Theta_B}}, \tag{97}$$

which is not approximately proportional to any expression of the form (96).

Before proceeding farther in the investigation of the distribution in phase which we have called canonical, it will be interesting to see whether the properties with respect to

* It will be observed that the above condition relating to the forces which act between the different systems is entirely analogous to that which must hold in the corresponding case in thermodynamics. The most simple test of the equality of temperature of two bodies is that they remain in equilibrium when brought into thermal contact. Direct thermal contact implies molecular forces acting between the bodies. Now the test will fail unless the energy of these forces can be neglected in comparison with the other energies of the bodies. Thus, in the case of energetic chemical action between the bodies, or when the number of particles affected by the forces acting between the bodies is not negligible in comparison with the whole number of particles (as when the bodies have the form of exceedingly thin sheets), the contact of bodies of the same temperature may produce considerable thermal disturbance, and thus fail to afford a reliable criterion of the equality of temperature.

statistical equilibrium which have been described are peculiar to it, or whether other distributions may have analogous properties.

Let η' and η'' be the indices of probability in two independent ensembles which are each in statistical equilibrium, then $\eta' + \eta''$ will be the index in the ensemble obtained by combining each system of the first ensemble with each system of the second. This third ensemble will of course be in statistical equilibrium, and the function of phase $\eta' + \eta''$ will be a constant of motion. Now when infinitesimal forces are added to the compound systems, if $\eta' + \eta''$ or a function differing infinitesimally from this is still a constant of motion, it must be on account of the nature of the forces added, or if their action is not entirely specified, on account of conditions to which they are subject. Thus, in the case already considered, $\eta' + \eta''$ is a function of the energy of the compound system, and the infinitesimal forces added are subject to the law of conservation of energy.

Another natural supposition in regard to the added forces is that they should be such as not to affect the moments of momentum of the compound system. To get a case in which moments of momentum of the compound system shall be constants of motion, we may imagine material particles contained in two concentric spherical shells, being prevented from passing the surfaces bounding the shells by repulsions acting always in lines passing through the common centre of the shells. Then, if there are no forces acting between particles in different shells, the mass of particles in each shell will have, besides its energy, the moments of momentum about three axes through the centre as constants of motion.

Now let us imagine an ensemble formed by distributing in phase the system of particles in one shell according to the index of probability

$$A - \frac{\epsilon}{\Theta} + \frac{\omega_1}{\Omega_1} + \frac{\omega_2}{\Omega_2} + \frac{\omega_3}{\Omega_3}, \tag{98}$$

where ϵ denotes the energy of the system, and ω_1, ω_2, ω_3, its three moments of momentum, and the other letters constants.

In like manner let us imagine a second ensemble formed by distributing in phase the system of particles in the other shell according to the index

$$A' - \frac{\epsilon'}{\Theta} + \frac{\omega_1'}{\Omega_1} + \frac{\omega_2'}{\Omega_2} + \frac{\omega_3'}{\Omega_3}, \tag{99}$$

where the letters have similar significations, and $\Theta, \Omega_1, \Omega_2, \Omega_3$ the same values as in the preceding formula. Each of the two ensembles will evidently be in statistical equilibrium, and therefore also the ensemble of compound systems obtained by combining each system of the first ensemble with each of the second. In this third ensemble the index of probability will be

$$A + A' - \frac{\epsilon + \epsilon'}{\Theta} + \frac{\omega_1 + \omega_1'}{\Omega_1} + \frac{\omega_2 + \omega_2'}{\Omega_2} + \frac{\omega_3 + \omega_3'}{\Omega_3}, \tag{100}$$

where the four numerators represent functions of phase which are constants of motion for the compound systems.

Now if we add in each system of this third ensemble infinitesimal conservative forces of attraction or repulsion between particles in different shells, determined by the same law for all the systems, the functions $\omega_1 + \omega'$, $\omega_2 + \omega_2'$, and $\omega_3 + \omega_3'$ will remain constants of motion, and a function differing infinitely little from $\epsilon_1 + \epsilon'$ will be a constant of motion. It would therefore require only an infinitesimal change in the distribution in phase of the ensemble of compound systems to make it a case of statistical equilibrium. These properties are entirely analogous to those of canonical ensembles.*

Again, if the relations between the forces and the coördinates can be expressed by linear equations, there will be certain "normal" types of vibration of which the actual motion may be regarded as composed, and the whole energy may be divided

* It would not be possible to omit the term relating to energy in the above indices, since without this term the condition expressed by equation (89) cannot be satisfied.

The consideration of the above case of statistical equilibrium may be made the foundation of the theory of the thermodynamic equilibrium of rotating bodies, — a subject which has been treated by Maxwell in his memoir "On Boltzmann's theorem on the average distribution of energy in a system of material points." Cambr. Phil. Trans., vol. XII, p. 547, (1878).

into parts relating separately to vibrations of these different types. These partial energies will be constants of motion, and if such a system is distributed according to an index which is any function of the partial energies, the ensemble will be in statistical equilibrium. Let the index be a linear function of the partial energies, say

$$A - \frac{\epsilon_1}{\Theta_1} \dots - \frac{\epsilon_n}{\Theta_n}. \qquad (101)$$

Let us suppose that we have also a second ensemble composed of systems in which the forces are linear functions of the coördinates, and distributed in phase according to an index which is a linear function of the partial energies relating to the normal types of vibration, say

$$A' - \frac{\epsilon_1'}{\Theta_1'} \dots - \frac{\epsilon_m'}{\Theta_m'}. \qquad (102)$$

Since the two ensembles are both in statistical equilibrium, the ensemble formed by combining each system of the first with each system of the second will also be in statistical equilibrium. Its distribution in phase will be represented by the index

$$A + A' - \frac{\epsilon_1}{\Theta_1} \dots - \frac{\epsilon_n}{\Theta_n} - \frac{\epsilon_1'}{\Theta_1'} \dots - \frac{\epsilon_m'}{\Theta_m'}, \qquad (103)$$

and the partial energies represented by the numerators in the formula will be constants of motion of the compound systems which form this third ensemble.

Now if we add to these compound systems infinitesimal forces acting between the component systems and subject to the same general law as those already existing, viz., that they are conservative and linear functions of the coördinates, there will still be $n + m$ types of normal vibration, and $n + m$ partial energies which are independent constants of motion. If all the original $n + m$ normal types of vibration have different periods, the new types of normal vibration will differ infinitesimally from the old, and the new partial energies, which are constants of motion, will be nearly the same functions of phase as the old. Therefore the distribution in phase of the ensemble of compound systems after the addition of the supposed infinitesimal forces will differ infinitesimally from one which would be in statistical equilibrium.

The case is not so simple when some of the normal types of motion have the same periods. In this case the addition of infinitesimal forces may completely change the normal types of motion. But the sum of the partial energies for all the original types of vibration which have any same period, will be nearly identical (as a function of phase, *i. e.*, of the coördinates and momenta,) with the sum of the partial energies for the normal types of vibration which have the same, or nearly the same, period after the addition of the new forces. If, therefore, the partial energies in the indices of the first two ensembles (101) and (102) which relate to types of vibration having the same periods, have the same divisors, the same will be true of the index (103) of the ensemble of compound systems, and the distribution represented will differ infinitesimally from one which would be in statistical equilibrium after the addition of the new forces.*

The same would be true if in the indices of each of the original ensembles we should substitute for the term or terms relating to any period which does not occur in the other ensemble, any function of the total energy related to that period, subject only to the general limitation expressed by equation (89). But in order that the ensemble of compound systems (with the added forces) shall always be approximately in statistical equilibrium, it is necessary that the indices of the original ensembles should be linear functions of those partial energies which relate to vibrations of periods common to the two ensembles, and that the coefficients of such partial energies should be the same in the two indices.†

* It is interesting to compare the above relations with the laws respecting the exchange of energy between bodies by radiation, although the phenomena of radiations lie entirely without the scope of the present treatise, in which the discussion is limited to systems of a finite number of degrees of freedom.

† The above may perhaps be sufficiently illustrated by the simple case where $n = 1$ in each system. If the periods are different in the two systems, they may be distributed according to any functions of the energies: but if

The properties of canonically distributed ensembles of systems with respect to the equilibrium of the new ensembles which may be formed by combining each system of one ensemble with each system of another, are therefore not peculiar to them in the sense that analogous properties do not belong to some other distributions under special limitations in regard to the systems and forces considered. Yet the canonical distribution evidently constitutes the most simple case of the kind, and that for which the relations described hold with the least restrictions.

Returning to the case of the canonical distribution, we shall find other analogies with thermodynamic systems, if we suppose, as in the preceding chapters,* that the potential energy (ϵ_q) depends not only upon the coördinates $q_1 \ldots q_n$ which determine the configuration of the system, but also upon certain coördinates a_1, a_2, etc. of bodies which we call *external*, meaning by this simply that they are not to be regarded as forming any part of the system, although their positions affect the forces which act on the system. The forces exerted by the system upon these external bodies will be represented by $-d\epsilon_q/da_1$, $-d\epsilon_q/da_2$, etc., while $-d\epsilon_q/dq_1$, $\ldots -d\epsilon_q/dq_n$ represent all the forces acting upon the bodies of the system, including those which depend upon the position of the external bodies, as well as those which depend only upon the configuration of the system itself. It will be understood that ϵ_p depends only upon $q_1, \ldots q_n, p_1, \ldots p_n$, in other words, that the kinetic energy of the bodies which we call external forms no part of the kinetic energy of the system. It follows that we may write

$$\frac{d\epsilon}{da_1} = \frac{d\epsilon_q}{da_1} = -A_1, \tag{104}$$

although a similar equation would not hold for differentiations relative to the internal coördinates.

the periods are the same they must be distributed canonically with same modulus in order that the compound ensemble with additional forces may be in statistical equilibrium.

* See especially Chapter I, p. 4.

We always suppose these external coördinates to have the same values for all systems of any ensemble. In the case of a canonical distribution, *i. e.*, when the index of probability of phase is a linear function of the energy, it is evident that the values of the external coördinates will affect the distribution, since they affect the energy. In the equation

$$e^{-\frac{\psi}{\Theta}} = \int_{\text{phases}}^{\text{all}} \cdots \int e^{-\frac{\epsilon}{\Theta}} dp_1 \ldots dq_n, \tag{105}$$

by which ψ may be determined, the external coördinates, a_1, a_2, etc., contained implicitly in ϵ, as well as Θ, are to be regarded as constant in the integrations indicated. The equation indicates that ψ is a function of these constants. If we imagine their values varied, and the ensemble distributed canonically according to their new values, we have by differentiation of the equation

$$\begin{aligned} e^{-\frac{\psi}{\Theta}}\left(-\frac{1}{\Theta}d\psi + \frac{\psi}{\Theta^2}d\Theta\right) = \frac{1}{\Theta^2}d\Theta \int_{\text{phases}}^{\text{all}} \cdots \int \epsilon\, e^{-\frac{\epsilon}{\Theta}} dp_1 \ldots dq_n \\ - \frac{1}{\Theta}da_1 \int_{\text{phases}}^{\text{all}} \cdots \int \frac{d\epsilon}{da_1} e^{-\frac{\epsilon}{\Theta}} dp_1 \ldots dq_n \\ - \frac{1}{\Theta}da_2 \int_{\text{phases}}^{\text{all}} \cdots \int \frac{d\epsilon}{da_2} e^{-\frac{\epsilon}{\Theta}} dp_1 \ldots dq_n - \text{etc.}, \end{aligned} \tag{106}$$

or, multiplying by $\Theta\, e^{\frac{\psi}{\Theta}}$, and setting

$$-\frac{d\epsilon}{da_1} = A_1, \qquad -\frac{d\epsilon}{da_2} = A_2, \quad \text{etc.,}$$

$$\begin{aligned} -d\psi + \frac{\psi}{\Theta}d\Theta = \frac{1}{\Theta}d\Theta \int_{\text{phases}}^{\text{all}} \cdots \int \epsilon\, e^{\frac{\psi-\epsilon}{\Theta}} dp_1 \ldots dq_n \\ + da_1 \int_{\text{phases}}^{\text{all}} \cdots \int A_1\, e^{\frac{\psi-\epsilon}{\Theta}} dp_1 \ldots dq_n \\ + da_2 \int_{\text{phases}}^{\text{all}} \cdots \int A_2\, e^{\frac{\psi-\epsilon}{\Theta}} dp_1 \ldots dq_n + \text{etc.} \end{aligned} \tag{107}$$

Now the average value in the ensemble of any quantity (which we shall denote in general by a horizontal line above the proper symbol) is determined by the equation

$$\bar{u} = \int_{\text{phases}}^{\text{all}} \cdots \int u\, e^{\frac{\psi - \epsilon}{\Theta}}\, dp_1 \ldots dq_n. \tag{108}$$

Comparing this with the preceding equation, we have

$$d\psi = \frac{\psi}{\Theta}\, d\Theta - \frac{\bar{\epsilon}}{\Theta}\, d\Theta - \bar{A}_1\, da_1 - \bar{A}_2\, da_2 - \text{etc.} \tag{109}$$

Or, since

$$\frac{\psi - \epsilon}{\Theta} = \eta, \tag{110}$$

and

$$\frac{\psi - \bar{\epsilon}}{\Theta} = \bar{\eta}, \tag{111}$$

$$d\psi = \bar{\eta}\, d\Theta - \bar{A}_1\, da_1 - \bar{A}_2\, da_2 - \text{etc.} \tag{112}$$

Moreover, since (111) gives

$$d\psi - d\bar{\epsilon} = \Theta\, d\bar{\eta} + \bar{\eta}\, d\Theta, \tag{113}$$

we have also

$$d\bar{\epsilon} = -\Theta\, d\bar{\eta} - \bar{A}_1\, da_1 - \bar{A}_2\, da_2 - \text{etc.} \tag{114}$$

This equation, if we neglect the sign of averages, is identical in form with the thermodynamic equation

$$d\eta = \frac{d\epsilon + A_1\, da_1 + A_2\, da_2 + \text{etc.}}{T}, \tag{115}$$

or

$$d\epsilon = T\, d\eta - A_1\, da_1 - A_2\, da_2 - \text{etc.}, \tag{116}$$

which expresses the relation between the energy, temperature, and entropy of a body in thermodynamic equilibrium, and the forces which it exerts on external bodies, — a relation which is the mathematical expression of the second law of thermodynamics for reversible changes. The modulus in the statistical equation corresponds to temperature in the thermodynamic equation, and the average index of probability *with its sign reversed* corresponds to entropy. But in the thermodynamic equation the entropy (η) is a quantity which is only defined by the equation itself, and incompletely defined in that the equation only determines its differential, and the constant of integration is arbitrary. On the other hand, the $\bar{\eta}$ in the statistical equation has been completely defined as the average value in a canonical ensemble of systems of the logarithm of the coefficient of probability of phase.

We may also compare equation (112) with the thermodynamic equation

$$\psi = -\eta dT - A_1 da_1 - A_2 da_2 - \text{etc.}, \tag{117}$$

where ψ represents the function obtained by subtracting the product of the temperature and entropy from the energy.

How far, or in what sense, the similarity of these equations constitutes any demonstration of the thermodynamic equations, or accounts for the behavior of material systems, as described in the theorems of thermodynamics, is a question of which we shall postpone the consideration until we have further investigated the properties of an ensemble of systems distributed in phase according to the law which we are considering. The analogies which have been pointed out will at least supply the motive for this investigation, which will naturally commence with the determination of the average values in the ensemble of the most important quantities relating to the systems, and to the distribution of the ensemble with respect to the different values of these quantities.

CHAPTER V.

AVERAGE VALUES IN A CANONICAL ENSEMBLE OF SYSTEMS.

In the simple but important case of a system of material points, if we use rectangular coördinates, we have for the product of the differentials of the coördinates

$$dx_1\, dy_1\, dz_1 \ldots dx_\nu\, dy_\nu\, dz_\nu,$$

and for the product of the differentials of the momenta

$$m_1\, d\dot{x}_1\, m_1\, d\dot{y}_1\, m_1\, d\dot{z}_1 \ldots m_\nu\, d\dot{x}_\nu\, m_\nu\, d\dot{y}_\nu\, m_\nu\, d\dot{z}_\nu.$$

The product of these expressions, which represents an element of extension-in-phase, may be briefly written

$$m_1\, d\dot{x}_1 \ldots m_\nu\, d\dot{z}_\nu\, dx_1 \ldots dz_\nu;$$

and the integral

$$\int \ldots \int e^{\frac{\psi - \epsilon}{\Theta}}\, m_1\, d\dot{x}_1 \ldots m_\nu\, d\dot{z}_\nu\, dx_1 \ldots dz_\nu \tag{118}$$

will represent the probability that a system taken at random from an ensemble canonically distributed will fall within any given limits of phase.

In this case

$$\epsilon = \epsilon_q + \tfrac{1}{2} m_1 \dot{x}_1^2 \ldots + \tfrac{1}{2} m_\nu \dot{z}_\nu^2, \tag{119}$$

and

$$e^{\frac{\psi - \epsilon}{\Theta}} = e^{\frac{\psi - \epsilon_q}{\Theta}}\, e^{-\frac{m_1 \dot{x}_1^2}{2\Theta}} \ldots e^{-\frac{m_\nu \dot{z}_\nu^2}{2\Theta}}. \tag{120}$$

The potential energy (ϵ_q) is independent of the velocities, and if the limits of integration for the coördinates are independent of the velocities, and the limits of the several velocities are independent of each other as well as of the coördinates, the multiple integral may be resolved into the product of integrals

$$\int \cdots \int e^{\frac{\psi - \epsilon_q}{\Theta}} dx_1 \ldots dz_\nu \int e^{-\frac{m_1 \dot{x}_1^2}{2\Theta}} m_1 d\dot{x}_1 \ldots \int e^{-\frac{m_\nu \dot{z}_\nu^2}{2\Theta}} m_\nu d\dot{z}_\nu. \quad (121)$$

This shows that the probability that the configuration lies within any given limits is independent of the velocities, and that the probability that any component velocity lies within any given limits is independent of the other component velocities and of the configuration.

Since

$$\int_{-\infty}^{+\infty} e^{-\frac{m_1 \dot{x}_1^2}{2\Theta}} m_1 d\dot{x}_1 = \sqrt{2\pi m_1 \Theta}, \quad (122)$$

and

$$\int_{-\infty}^{+\infty} \tfrac{1}{2} m_1 \dot{x}_1^2 e^{-\frac{m_1 \dot{x}_1^2}{2\Theta}} m_1 d\dot{x}_1 = \sqrt{\tfrac{1}{2}\pi m_1 \Theta^3}, \quad (123)$$

the average value of the part of the kinetic energy due to the velocity $\dot{x}_1$, which is expressed by the quotient of these integrals, is $\frac{1}{2}\Theta$. This is true whether the average is taken for the whole ensemble or for any particular configuration, whether it is taken without reference to the other component velocities, or only those systems are considered in which the other component velocities have particular values or lie within specified limits.

The number of coördinates is 3ν or n. We have, therefore, for the average value of the kinetic energy of a system

$$\bar{\epsilon}_p = \tfrac{3}{2}\nu\Theta = \tfrac{1}{2}n\Theta. \quad (124)$$

This is equally true whether we take the average for the whole ensemble, or limit the average to a single configuration.

The distribution of the systems with respect to their component velocities follows the 'law of errors'; the probability that the value of any component velocity lies within any given limits being represented by the value of the corresponding integral in (121) for those limits, divided by $(2\pi m\Theta)^{\frac{1}{2}}$, which is the value of the same integral for infinite limits. Thus the probability that the value of $\dot{x}_1$ lies between any given limits is expressed by

$$\left(\frac{m_1}{2\pi\Theta}\right)^{\frac{1}{2}} \int e^{-\frac{m_1 \dot{x}_1^2}{2\Theta}} d\dot{x}_1. \tag{125}$$

The expression becomes more simple when the velocity is expressed with reference to the energy involved. If we set

$$s = \left(\frac{m_1}{2\Theta}\right)^{\frac{1}{2}} \dot{x}_1,$$

the probability that s lies between any given limits is expressed by

$$\frac{1}{\sqrt{\pi}} \int e^{-s^2} ds. \tag{126}$$

Here s is the ratio of the component velocity to that which would give the energy Θ; in other words, s^2 is the quotient of the energy due to the component velocity divided by Θ. The distribution with respect to the partial energies due to the component velocities is therefore the same for all the component velocities.

The probability that the configuration lies within any given limits is expressed by the value of

$$M^{\frac{1}{2}} (2\pi\Theta)^{\frac{3\nu}{2}} \int \dots \int e^{\frac{\psi - \epsilon_q}{\Theta}} dx_1 \dots dz_\nu \tag{127}$$

for those limits, where M denotes the product of all the masses. This is derived from (121) by substitution of the values of the integrals relating to velocities taken for infinite limits.

Very similar results may be obtained in the general case of a conservative system of n degrees of freedom. Since ϵ_p is a homogeneous quadratic function of the p's, it may be divided into parts by the formula

$$\epsilon_p = \tfrac{1}{2} p_1 \frac{d\epsilon_p}{dp_1} \dots + \tfrac{1}{2} p_n \frac{d\epsilon_p}{dp_n} \tag{128}$$

where ϵ might be written for ϵ_p in the differential coefficients without affecting the signification. The average value of the first of these parts, for any given configuration, is expressed by the quotient

$$\frac{\int_{-\infty}^{+\infty} \cdots \int_{-\infty}^{+\infty} \tfrac{1}{2} p_1 \frac{d\epsilon}{dp_1} e^{\frac{\psi - \epsilon}{\Theta}} dp_1 \ldots dp_n}{\int_{-\infty}^{+\infty} \cdots \int_{-\infty}^{+\infty} e^{\frac{\psi - \epsilon}{\Theta}} dp_1 \ldots dp_n} \tag{129}$$

Now we have by integration by parts

$$\int_{-\infty}^{+\infty} p_1 e^{\frac{\psi - \epsilon}{\Theta}} \frac{d\epsilon}{dp_1} dp_1 = \Theta \int_{-\infty}^{+\infty} e^{\frac{\psi - \epsilon}{\Theta}} dp_1. \tag{130}$$

By substitution of this value, the above quotient reduces to $\frac{\Theta}{2}$, which is therefore the average value of $\tfrac{1}{2} p_1 \frac{d\epsilon}{dp_1}$ for the given configuration. Since this value is independent of the configuration, it must also be the average for the whole ensemble, as might easily be proved directly. (To make the preceding proof apply directly to the whole ensemble, we have only to write $dp_1 \ldots dq_n$ for $dp_1 \ldots dp_n$ in the multiple integrals.) This gives $\tfrac{1}{2} n \Theta$ for the average value of the whole kinetic energy for any given configuration, or for the whole ensemble, as has already been proved in the case of material points.

The mechanical significance of the several parts into which the kinetic energy is divided in equation (128) will be apparent if we imagine that by the application of suitable forces (different from those derived from ϵ_q and so much greater that the latter may be neglected in comparison) the system was brought from rest to the state of motion considered, so rapidly that the configuration was not sensibly altered during the process, and in such a manner also that the ratios of the component velocities were constant in the process. If we write

$$F_1 dq_1 \ldots + F_n dq_n$$

for the moment of these forces, we have for the period of their action by equation (3)

$$\dot{p}_1 = -\frac{d\epsilon_p}{dq_1} - \frac{d\epsilon_q}{dq_1} + F_1 = -\frac{d\epsilon}{dq_1} + F_1.$$

The work done by the force F_1 may be evaluated as follows:

$$\int F_1\, dq_1 = \int \dot{p}_1\, dq_1 + \int \frac{d\epsilon}{dq_1}\, dq_1,$$

where the last term may be cancelled because the configuration does not vary sensibly during the application of the forces. (It will be observed that the other terms contain factors which increase as the time of the action of the forces is diminished.) We have therefore,

$$\int F_1\, dq_1 = \int \dot{p}_1 \dot{q}_1\, dt = \int \dot{q}_1\, dp = \frac{\dot{q}_1}{p_1} \int p_1\, dp_1. \tag{131}$$

For since the p's are linear functions of the $\dot{q}$'s (with coefficients involving the q's) the supposed constancy of the q's and of the ratios of the $\dot{q}$'s will make the ratio $\dot{q}_1 / p_1$ constant. The last integral is evidently to be taken between the limits zero and the value of p_1 in the phase originally considered, and the quantities before the integral sign may be taken as relating to that phase. We have therefore

$$\int F_1\, dq_1 = \tfrac{1}{2} p_1 \dot{q}_1 = \tfrac{1}{2} p_1 \frac{d\epsilon}{dp_1}. \tag{132}$$

That is: the several parts into which the kinetic energy is divided in equation (128) represent the amounts of energy communicated to the system by the several forces $F_1, \ldots F_n$ under the conditions mentioned.

The following transformation will not only give the value of the average kinetic energy, but will also serve to separate the distribution of the ensemble in configuration from its distribution in velocity.

Since $2\,\epsilon_p$ is a homogeneous quadratic function of the p's, which is incapable of a negative value, it can always be expressed (and in more than one way) as a sum of squares of linear functions of the p's.* The coefficients in these linear functions, like those in the quadratic function, must be regarded in the general case as functions of the q's. Let

$$2\,\epsilon_p = u_1^2 + u_2^2 \ldots + u_n^2 \tag{133}$$

where $u_1 \ldots u_n$ are such linear functions of the p's. If we write

$$\frac{d(p_1 \ldots p_n)}{d(u_1 \ldots u_n)}$$

for the Jacobian or determinant of the differential coefficients of the form dp/du, we may substitute

$$\frac{d(p_1 \ldots p_n)}{d(u_1 \ldots u_n)}\, du_1 \ldots du_n$$

for

$$dp_1 \ldots dp_n$$

under the multiple integral sign in any of our formulæ. It will be observed that this determinant is function of the q's alone. The sign of such a determinant depends on the relative order of the variables in the numerator and denominator. But since the suffixes of the u's are only used to distinguish these functions from one another, and no especial relation is supposed between a p and a u which have the same suffix, we may evidently, without loss of generality, suppose the suffixes so applied that the determinant is positive.

Since the u's are linear functions of the p's, when the integrations are to cover all values of the p's (for constant q's) once and only once, they must cover all values of the u's once and only once, and the limits will be $\pm\infty$ for all the u's. Without the supposition of the last paragraph the upper limits would not always be $+\infty$, as is evident on considering the effect of changing the sign of a u. But with the supposition which we have made (that the determinant is always positive) we may make the upper limits $+\infty$ and the lower $-\infty$ for all the u's. Analogous considerations will apply where the integrations do not cover all values of the p's and therefore of

* The reduction requires only the repeated application of the process of 'completing the square' used in the solution of quadratic equations.

the u's. The integrals may always be taken from a less to a greater value of a u.

The general integral which expresses the fractional part of the ensemble which falls within any given limits of phase is thus reduced to the form

$$\int \cdots \int e^{\frac{\psi - \epsilon_q}{\Theta}} \frac{d(p_1 \ldots p_n)}{d(u_1 \ldots u_n)} e^{-\frac{u_1^2 \ldots u_n^2}{2\Theta}} du_1 \ldots du_n \, dq_1 \ldots dq_n. \tag{134}$$

For the average value of the part of the kinetic energy which is represented by $\frac{1}{2} u_1^2$, whether the average is taken for the whole ensemble, or for a given configuration, we have therefore

$$\tfrac{1}{2}\overline{u_1^2} = \frac{\displaystyle\int_{-\infty}^{+\infty} \tfrac{1}{2} u_1^2 e^{-\frac{u_1^2}{2\Theta}} du_1}{\displaystyle\int_{-\infty}^{+\infty} e^{-\frac{u_1^2}{2\Theta}} du_1} = \frac{(\frac{1}{2}\pi\Theta^3)^{\frac{1}{2}}}{(2\pi\Theta)^{\frac{1}{2}}} = \frac{\Theta}{2}, \tag{135}$$

and for the average of the whole kinetic energy, $\frac{1}{2} n \Theta$, as before.

The fractional part of the ensemble which lies within any given limits of *configuration*, is found by integrating (134) with respect to the u's from $-\infty$ to $+\infty$. This gives

$$(2\pi\Theta)^{\frac{n}{2}} \int \cdots \int e^{\frac{\psi - \epsilon_q}{\Theta}} \frac{d(p_1 \ldots p_n)}{d(u_1 \ldots u_n)} dq_1 \ldots dq_n \tag{136}$$

which shows that the value of the Jacobian is independent of the manner in which $2\,\epsilon_p$ is divided into a sum of squares. We may verify this directly, and at the same time obtain a more convenient expression for the Jacobian, as follows.

It will be observed that since the u's are linear functions of the p's, and the p's linear functions of the $\dot{q}$'s, the u's will be linear functions of the $\dot{q}$'s, so that a differential coefficient of the form $du/d\dot{q}$ will be independent of the $\dot{q}$'s, and function of the q's alone. Let us write dp_x/du_y for the general element of the Jacobian determinant. We have

$$\frac{dp_x}{du_y} = \frac{d}{du_y}\frac{d\epsilon}{d\dot{q}_x} = \frac{d}{du_y}\sum_{r=1}^{r=n}\frac{d\epsilon}{du_r}\frac{du_r}{d\dot{q}_x}$$

$$= \sum_{r=1}^{r=n}\left(\frac{d^2\epsilon}{du_y\,du_r}\frac{du_r}{d\dot{q}_x}\right) = \frac{d}{d\dot{q}_x}\frac{d\epsilon}{du_y} = \frac{du_y}{d\dot{q}_x} \tag{137}$$

Therefore

$$\frac{d(p, \ldots p_n)}{d(u, \ldots u_n)} = \frac{d(u, \ldots u_n)}{d(\dot{q}, \ldots \dot{q}_n)} \tag{138}$$

and

$$\left(\frac{d(p, \ldots p_n)}{d(u, \ldots u_n)}\right)^2 = \left(\frac{d(u, \ldots u_n)}{d(\dot{q}, \ldots \dot{q}_n)}\right)^2 = \frac{d(p, \ldots p_n)}{d(\dot{q}, \ldots \dot{q}_n)} \tag{139}$$

These determinants are all functions of the q's alone.* The last is evidently the Hessian or determinant formed of the second differential coefficients of the kinetic energy with respect to $\dot{q}_1, \ldots \dot{q}_n$. We shall denote it by $\Delta_{\dot{q}}$. The reciprocal determinant

$$\frac{d(\dot{q}_1 \ldots \dot{q}_n)}{d(p_1 \ldots p_n)},$$

which is the Hessian of the kinetic energy regarded as function of the p's, we shall denote by Δ_p.

If we set

$$e^{-\frac{\psi_p}{\Theta}} = \int_{-\infty}^{+\infty} \ldots \int_{-\infty}^{+\infty} e^{-\frac{\epsilon_p}{\Theta}} \Delta_p^{\frac{1}{2}}\, dp, \ldots dp_n$$

$$= \int_{-\infty}^{+\infty} \ldots \int_{-\infty}^{+\infty} e^{\frac{-u_1^2 \ldots -u_n^2}{2\Theta}} du_1 \ldots du_n = (2\pi\Theta)^{\frac{n}{2}}, \tag{140}$$

and

$$\psi_q = \psi - \psi_p, \tag{141}$$

* It will be observed that the proof of (137) depends on the *linear* relation between the u's and $\dot{q}$'s, which makes $\frac{du_r}{d\dot{q}_x}$ constant with respect to the differentiations here considered. Compare note on p. 12.

the fractional part of the ensemble which lies within any given limits of configuration (136) may be written

$$\int\cdots\int e^{\frac{\psi_q-\epsilon_q}{\Theta}}\,\Delta_{\dot q}^{\frac{1}{2}}\,dq_1\ldots dq_n, \qquad (142)$$

where the constant ψ_q may be determined by the condition that the integral extended over all configurations has the value unity.*

* In the simple but important case in which $\Delta_{\dot q}$ is independent of the q's, and ϵ_q a quadratic function of the q's, if we write ϵ_a for the least value of ϵ_q (or of ϵ) consistent with the given values of the external coördinates, the equation determining ψ_q may be written

$$e^{\frac{\epsilon_a-\psi_q}{\Theta}} = \Delta_{\dot q}^{\frac{1}{2}}\int_{-\infty}^{+\infty}\cdots\int_{-\infty}^{+\infty} e^{-\frac{(\epsilon_q-\epsilon_a)}{\Theta}}\,dq_1\ldots dq_n.$$

If we denote by $q_1', \ldots q_n'$ the values of $q_1, \ldots q_n$ which give ϵ_q its least value ϵ_a, it is evident that $\epsilon_q-\epsilon_a$ is a homogenous quadratic function of the differences q_1-q_1', etc., and that $dq_1, \ldots dq_n$ may be regarded as the differentials of these differences. The evaluation of this integral is therefore analytically similar to that of the integral

$$\int_{-\infty}^{+\infty}\cdots\int_{-\infty}^{+\infty} e^{-\frac{\epsilon_p}{\Theta}}\,dp_1\ldots dp_n,$$

for which we have found the value $\Delta_p^{-\frac{1}{2}}(2\pi\Theta)^{\frac{n}{2}}$. By the same method, or by analogy, we get

$$e^{\frac{\epsilon_a-\psi_q}{\Theta}} = \left(\frac{\Delta_{\dot q}}{\Delta_q}\right)^{\frac{1}{2}}(2\pi\Theta)^{\frac{n}{2}},$$

where Δ_q is the Hessian of the potential energy as function of the q's. It will be observed that Δ_q depends on the forces of the system and is independent of the masses, while $\Delta_{\dot q}$ or its reciprocal Δ_p depends on the masses and is independent of the forces. While each Hessian depends on the system of coördinates employed, the ratio $\Delta_q/\Delta_{\dot q}$ is the same for all systems.

Multiplying the last equation by (140), we have

$$e^{\frac{\epsilon_a-\psi}{\Theta}} = \left(\frac{\Delta_{\dot q}}{\Delta_q}\right)^{\frac{1}{2}}(2\pi\Theta)^{n}.$$

For the average value of the potential energy, we have

$$\overline{\epsilon_q}-\epsilon_a = \frac{\displaystyle\int_{-\infty}^{+\infty}\cdots\int_{-\infty}^{+\infty}(\epsilon_q-\epsilon_a)\,e^{-\frac{\epsilon_q-\epsilon_a}{\Theta}}\,dq_1\ldots dq_n}{\displaystyle\int_{-\infty}^{+\infty}\cdots\int_{-\infty}^{+\infty} e^{-\frac{\epsilon_q-\epsilon_a}{\Theta}}\,dq_1\ldots dq_n}.$$

When an ensemble of systems is distributed in *configuration* in the manner indicated in this formula, *i. e.*, when its distribution in configuration is the same as that of an ensemble canonically distributed in phase, we shall say, without any reference to its velocities, that it is *canonically distributed in configuration.*

For any given configuration, the fractional part of the systems which lie within any given limits of velocity is represented by the quotient of the multiple integral

$$\int \cdots \int e^{-\frac{\epsilon_p}{\Theta}} dp_1 \ldots dp_n,$$

or its equivalent

$$\int \cdots \int e^{\frac{-u_1^2 \ldots -u_n^2}{2\Theta}} \Delta_q^{\frac{1}{2}} du_1 \ldots du_n,$$

taken within those limits divided by the value of the same integral for the limits $\pm\infty$. But the value of the second multiple integral for the limits $\pm\infty$ is evidently

$$\Delta_q^{\frac{1}{2}} (2\pi\Theta)^{\frac{n}{2}}.$$

We may therefore write

$$\int \cdots \int e^{\frac{\psi_p - \epsilon_p}{\Theta}} du_1 \ldots du_n, \tag{143}$$

The evaluation of this expression is similar to that of

$$\frac{\int\limits_{-\infty}^{+\infty} \cdots \int\limits_{-\infty}^{+\infty} \epsilon_p e^{-\frac{\epsilon_p}{\Theta}} dp_1 \ldots dp_n}{\int\limits_{-\infty}^{+\infty} \cdots \int\limits_{-\infty}^{+\infty} e^{-\frac{\epsilon_p}{\Theta}} dp_1 \ldots dp_n}$$

which expresses the average value of the kinetic energy, and for which we have found the value $\frac{1}{2} n\Theta$. We have accordingly

$$\overline{\epsilon_q} - \epsilon_a = \frac{1}{2} n\Theta.$$

Adding the equation

$$\overline{\epsilon_p} = \frac{1}{2} n\Theta,$$

we have

$$\overline{\epsilon} - \epsilon_a = n\Theta.$$

or
$$\int \dots \int e^{\frac{\psi_p - \epsilon_p}{\Theta}} \Delta_p^{\frac{1}{2}} dp_1 \dots dp_n, \tag{144}$$

or again
$$\int \dots \int e^{\frac{\psi_p - \epsilon_p}{\Theta}} \Delta_{\dot{q}}^{\frac{1}{2}} d\dot{q}_1 \dots d\dot{q}_n, \tag{145}$$

for the fractional part of the systems of any given configuration which lie within given limits of velocity.

When systems are distributed in velocity according to these formulae, *i. e.*, when the distribution in velocity is like that in an ensemble which is canonically distributed in phase, we shall say that they are *canonically distributed in velocity.*

The fractional part of the whole ensemble which falls within any given limits of phase, which we have before expressed in the form

$$\int \dots \int e^{\frac{\psi - \epsilon}{\Theta}} dp_1 \dots dp_n dq_1 \dots dq_n, \tag{146}$$

may also be expressed in the form

$$\int \dots \int e^{\frac{\psi - \epsilon}{\Theta}} \Delta_{\dot{q}} d\dot{q}_1 \dots d\dot{q}_n dq_1 \dots dq_n. \tag{147}$$

CHAPTER VI.

EXTENSION IN CONFIGURATION AND EXTENSION IN VELOCITY.

THE formulae relating to canonical ensembles in the closing paragraphs of the last chapter suggest certain general notions and principles, which we shall consider in this chapter, and which are not at all limited in their application to the canonical law of distribution.*

We have seen in Chapter IV. that the nature of the distribution which we have called canonical is independent of the system of coördinates by which it is described, being determined entirely by the modulus. It follows that the value represented by the multiple integral (142), which is the fractional part of the ensemble which lies within certain limiting configurations, is independent of the system of coördinates, being determined entirely by the limiting configurations with the modulus. Now ψ, as we have already seen, represents a value which is independent of the system of coördinates by which it is defined. The same is evidently true of ψ_p by equation (140), and therefore, by (141), of ψ_q. Hence the exponential factor in the multiple integral (142) represents a value which is independent of the system of coördinates. It follows that the value of a multiple integral of the form

$$\int \cdots \int \Delta_q^{\frac{1}{2}} \, dq_1 \ldots dq_n \qquad (148)$$

* These notions and principles are in fact such as a more logical arrangement of the subject would place in connection with those of Chapter I., to which they are closely related. The strict requirements of logical order have been sacrificed to the natural development of the subject, and very elementary notions have been left until they have presented themselves in the study of the leading problems.

is independent of the system of coördinates which is employed for its evaluation, as will appear at once, if we suppose the multiple integral to be broken up into parts so small that the exponential factor may be regarded as constant in each.

In the same way the formulae (144) and (145) which express the probability that a system (in a canonical ensemble) of given configuration will fall within certain limits of velocity, show that multiple integrals of the form

$$\int \cdots \int \Delta_p^{\frac{1}{2}}\, dp_1 \ldots dp_n \tag{149}$$

or

$$\int \cdots \int \Delta_{\dot{q}}^{\frac{1}{2}}\, d\dot{q}_1 \ldots d\dot{q}_n \tag{150}$$

relating to velocities possible for a given configuration, when the limits are formed by given velocities, have values independent of the system of coördinates employed.

These relations may easily be verified directly. It has already been proved that

$$\frac{d(P_1, \ldots P_n)}{d(p_1, \ldots p_n)} = \frac{d(\dot{q}_1 \ldots \dot{q}_n)}{d(\dot{Q}_1, \ldots \dot{Q}_n)} = \frac{d(q_1, \ldots q_n)}{d(Q_1, \ldots Q_n)}$$

where $q_1, \ldots q_n, p_1, \ldots p_n$ and $Q_1, \ldots Q_n, P_1, \ldots P_n$ are two systems of coördinates and momenta.* It follows that

$$\int \cdots \int \left(\frac{d(p_1, \ldots p_n)}{d(\dot{q}_1, \ldots \dot{q}_n)} \right)^{\frac{1}{2}} dq_1 \ldots dq_n$$

$$= \int \cdots \int \left(\frac{d(p_1, \ldots p_n)}{d(\dot{q}_1, \ldots \dot{q}_n)} \right)^{\frac{1}{2}} \frac{d(q_1, \ldots q_n)}{d(Q_1, \ldots Q_n)}\, dQ_1 \ldots dQ_n$$

$$= \int \cdots \int \left(\frac{d(p_1, \ldots p_n)}{d(\dot{q}_1, \ldots \dot{q}_n)} \right)^{\frac{1}{2}} \left(\frac{d(P_1, \ldots P_n)}{d(p_1, \ldots p_n)} \right)^{\frac{1}{2}} \left(\frac{d(\dot{q}_1, \ldots \dot{q}_n)}{d(\dot{Q}_1, \ldots \dot{Q}_n)} \right)^{\frac{1}{2}} dQ_1 \ldots dQ_n$$

$$= \int \cdots \int \left(\frac{d(P_1, \ldots P_n)}{d(\dot{Q}_1, \ldots \dot{Q}_n)} \right)^{\frac{1}{2}} dQ_1 \ldots dQ_n,$$

* See equation (29).

and

$$\int \cdots \int \left(\frac{d(\dot{Q}_1, \ldots \dot{Q}_n)}{d(P_1, \ldots P_n)} \right)^{\frac{1}{2}} dP_1 \ldots P_n$$

$$= \int \cdots \int \left(\frac{d(\dot{Q}_1, \ldots \dot{Q}_n)}{d(P_1, \ldots P_n)} \right)^{\frac{1}{2}} \frac{d(P_1, \ldots P_n)}{d(p_1, \ldots p_n)} dp_1 \ldots dp_n$$

$$= \int \cdots \int \left(\frac{d(\dot{Q}_1, \ldots \dot{Q}_n)}{d(P_1, \ldots P_n)} \right)^{\frac{1}{2}} \left(\frac{d(P_1, \ldots P_n)}{d(p_1, \ldots p_n)} \right)^{\frac{1}{2}} \left(\frac{d(\dot{q}_1, \ldots \dot{q}_n)}{d(\dot{Q}_1, \ldots \dot{Q}_n)} \right)^{\frac{1}{2}} dp_1 \ldots dp_n$$

$$= \int \cdots \int \left(\frac{d(\dot{q}_1, \ldots \dot{q}_n)}{d(p_1, \ldots p_n)} \right)^{\frac{1}{2}} dp_1 \ldots dp_n.$$

The multiple integral

$$\int \cdots \int dp_1 \ldots dp_n dq_1 \ldots dq_n, \tag{151}$$

which may also be written

$$\int \cdots \int \Delta_{\dot{q}} d\dot{q}_1 \ldots d\dot{q}_n dq_1 \ldots dq_n, \tag{152}$$

and which, when taken within any given limits of phase, has been shown to have a value independent of the coördinates employed, expresses what we have called an *extension-in-phase.** In like manner we may say that the multiple integral (148) expresses an *extension-in-configuration*, and that the multiple integrals (149) and (150) express an *extension-in-velocity*. We have called

$$dp_1 \ldots dp_n dq_1 \ldots dq_n, \tag{153}$$

which is equivalent to

$$\Delta_{\dot{q}} d\dot{q}_1 \ldots d\dot{q}_n dq_1 \ldots dq_n, \tag{154}$$

an element of extension-in-phase. We may call

$$\Delta_{\dot{q}}^{\frac{1}{2}} dq_1 \ldots dq_n \tag{155}$$

an element of extension-in-configuration, and

$$\Delta_p^{\frac{1}{2}} dp_1 \ldots dp_n, \tag{156}$$

* See Chapter I, p. 10.

or its equivalent

$$\Delta_{\dot{q}}^{\frac{1}{2}} d\dot{q}_1 \ldots d\dot{q}_n, \tag{157}$$

an element of extension-in-velocity.

An extension-in-phase may always be regarded as an integral of elementary extensions-in-configuration multiplied each by an extension-in-velocity. This is evident from the formulae (151) and (152) which express an extension-in-phase, if we imagine the integrations relative to velocity to be first carried out.

The product of the two expressions for an element of extension-in-velocity (149) and (150) is evidently of the same dimensions as the product

$$p_1 \ldots p_n \dot{q}_1 \ldots \dot{q}_n$$

that is, as the nth power of energy, since every product of the form $p_1 \dot{q}_1$ has the dimensions of energy. Therefore an extension-in-velocity has the dimensions of the square root of the nth power of energy. Again we see by (155) and (156) that the product of an extension-in-configuration and an extension-in-velocity have the dimensions of the nth power of energy multiplied by the nth power of time. Therefore an extension-in-configuration has the dimensions of the nth power of time multiplied by the square root of the nth power of energy.

To the notion of extension-in-configuration there attach themselves certain other notions analogous to those which have presented themselves in connection with the notion of extension-in-phase. The number of systems of any ensemble (whether distributed canonically or in any other manner) which are contained in an element of extension-in-configuration, divided by the numerical value of that element, may be called the *density-in-configuration.* That is, if a certain configuration is specified by the coördinates $q_1 \ldots q_n$, and the number of systems of which the coördinates fall between the limits q_1 and $q_1 + dq_1, \ldots q_n$ and $q_n + dq_n$ is expressed by

$$D_q \Delta_q^{\frac{1}{2}} dq_1 \ldots dq_n, \tag{158}$$

D_q will be the density-in-configuration. And if we set

$$e^{\eta_q} = \frac{D_q}{N}, \tag{159}$$

where N denotes, as usual, the total number of systems in the ensemble, the probability that an unspecified system of the ensemble will fall within the given limits of configuration, is expressed by

$$e^{\eta_q} \Delta_q^{\frac{1}{2}} dq_1 \ldots dq_n. \tag{160}$$

We may call e^{η_q} the *coefficient of probability of the configuration*, and η_q the *index of probability of the configuration.*

The fractional part of the whole number of systems which are within any given limits of configuration will be expressed by the multiple integral

$$\int \ldots \int e^{\eta_q} \Delta_q^{\frac{1}{2}} dq_1 \ldots dq_n. \tag{161}$$

The value of this integral (taken within any given configurations) is therefore independent of the system of coördinates which is used. Since the same has been proved of the same integral without the factor e^{η_q}, it follows that the values of η_q and D_q for a given configuration in a given ensemble are independent of the system of coördinates which is used.

The notion of extension-in-velocity relates to systems having the same configuration.* If an ensemble is distributed both in configuration and in velocity, we may confine our attention to those systems which are contained within certain infinitesimal limits of configuration, and compare the whole number of such systems with those which are also contained

* Except in some simple cases, such as a system of material points, we cannot compare velocities in one configuration with velocities in another, and speak of their identity or difference except in a sense entirely artificial. We may indeed say that we call the velocities in one configuration the same as those in another when the quantities $\dot{q}_1, \ldots \dot{q}_n$ have the same values in the two cases. But this signifies nothing until the system of coördinates has been defined. We might identify the velocities in the two cases which make the quantities $p_1, \ldots p_n$ the same in each. This again would signify nothing independently of the system of coördinates employed.

within certain infinitesimal limits of velocity. The second of these numbers divided by the first expresses the probability that a system which is only specified as falling within the infinitesimal limits of configuration shall also fall within the infinitesimal limits of velocity. If the limits with respect to velocity are expressed by the condition that the momenta shall fall between the limits p_1 and $p_1 + dp_1, \ldots p_n$ and $p_n + dp_n$, the extension-in-velocity within those limits will be

$$\Delta_p^{\frac{1}{2}} dp_1 \ldots dp_n,$$

and we may express the probability in question by

$$e^{\eta_p} \Delta_p^{\frac{1}{2}} dp_1 \ldots dp_n. \tag{162}$$

This may be regarded as defining η_p.

The probability that a system which is only specified as having a configuration within certain infinitesimal limits shall also fall within any given limits of velocity will be expressed by the multiple integral

$$\int \ldots \int e^{\eta_p} \Delta_p^{\frac{1}{2}} dp_1 \ldots dp_n, \tag{163}$$

or its equivalent

$$\int \ldots \int e^{\eta_p} \Delta_{\dot{q}}^{\frac{1}{2}} d\dot{q}_1 \ldots d\dot{q}_n, \tag{164}$$

taken within the given limits.

It follows that the probability that the system will fall within the limits of velocity, $\dot{q}_1$ and $\dot{q}_1 + d\dot{q}_1, \ldots \dot{q}_n$ and $\dot{q}_n + d\dot{q}_n$ is expressed by

$$e^{\eta_p} \Delta_{\dot{q}}^{\frac{1}{2}} d\dot{q}_1 \ldots d\dot{q}_n. \tag{165}$$

The value of the integrals (163), (164) is independent of the system of coördinates and momenta which is used, as is also the value of the same integrals without the factor e^{η_p}; therefore the value of η_p must be independent of the system of coördinates and momenta. We may call e^{η_p} the *coefficient of probability of velocity*, and η_p the *index of probability of velocity*.

Comparing (160) and (162) with (40), we get

$$e^{\eta_q} e^{\eta_p} = P = e^{\eta} \tag{166}$$

or

$$\eta_q + \eta_p = \eta. \tag{167}$$

That is: the product of the coefficients of probability of configuration and of velocity is equal to the coefficient of probability of phase; the sum of the indices of probability of configuration and of velocity is equal to the index of probability of phase.

It is evident that e^{η_q} and e^{η_p} have the dimensions of the reciprocals of extension-in-configuration and extension-in-velocity respectively, *i. e.*, the dimensions of $t^{-n}\,\epsilon^{-\frac{n}{2}}$ and $\epsilon^{-\frac{n}{2}}$, where t represent any time, and ϵ any energy. If, therefore, the unit of time is multiplied by c_t, and the unit of energy by c_ϵ, every η_q will be increased by the addition of

$$n \log c_t + \tfrac{1}{2} n \log c_\epsilon, \tag{168}$$

and every η_p by the addition of

$$\tfrac{1}{2} n \log c_\epsilon.* \tag{169}$$

It should be observed that the quantities which have been called *extension-in-configuration* and *extension-in-velocity* are not, as the terms might seem to imply, purely geometrical or kinematical conceptions. To express their nature more fully, they might appropriately have been called, respectively, *the dynamical measure of the extension in configuration*, and the *dynamical measure of the extension in velocity.* They depend upon the masses, although not upon the forces of the system. In the simple case of material points, where each point is limited to a given space, the extension-in-configuration is the product of the volumes within which the several points are confined (these may be the same or different), multiplied by the square root of the cube of the product of the masses of the several points. The extension-in-velocity for such systems is most easily defined as the extension-in-configuration of systems which have moved from the same configuration for the unit of time with the given velocities.

* Compare (47) in Chapter I.

In the general case, the notions of extension-in-configuration and extension-in-velocity may be connected as follows.

If an ensemble of similar systems of n degrees of freedom have the same configuration at a given instant, but are distributed throughout any finite extension-in-velocity, the same ensemble after an infinitesimal interval of time δt will be distributed throughout an extension in configuration equal to its original extension-in-velocity multiplied by δt^n.

In demonstrating this theorem, we shall write $q_1', \ldots q_n'$ for the initial values of the coördinates. The final values will evidently be connected with the initial by the equations

$$q_1 - q_1' = \dot{q}_1 \delta t, \ldots q_n - q_n' = \dot{q}_n \delta t. \tag{170}$$

Now the original extension-in-velocity is by definition represented by the integral

$$\int \ldots \int \Delta_q^{\frac{1}{2}} d\dot{q}_1 \ldots d\dot{q}_n, \tag{171}$$

where the limits may be expressed by an equation of the form

$$F(\dot{q}_1, \ldots \dot{q}_n) = 0. \tag{172}$$

The same integral multiplied by the constant δt^n may be written

$$\int \ldots \int \Delta_q^{\frac{1}{2}} d(\dot{q}_1 \delta t), \ldots d(\dot{q}_n \delta t), \tag{173}$$

and the limits may be written

$$F(\dot{q}_1 \ldots \dot{q}_n) = f(\dot{q}_1 \delta t, \ldots \dot{q}_n \delta t) = 0. \tag{174}$$

(It will be observed that δt as well as Δ_q is constant in the integrations.) Now this integral is identically equal to

$$\int \ldots \int \Delta_q^{\frac{1}{2}} d(q_1 - q_1') \ldots d(q_n \ldots q_n'), \tag{175}$$

or its equivalent

$$\int \ldots \int \Delta_q^{\frac{1}{2}} dq_1 \ldots dq_n, \tag{176}$$

with limits expressed by the equation

$$f(q_1 - q_1', \ldots q_n - q_n') = 0. \tag{177}$$

But the systems which initially had velocities satisfying the equation (172) will after the interval δt have configurations satisfying equation (177). Therefore the extension-in-configuration represented by the last integral is that which belongs to the systems which originally had the extension-in-velocity represented by the integral (171).

Since the quantities which we have called extensions-in-phase, extensions-in-configuration, and extensions-in-velocity are independent of the nature of the system of coördinates used in their definitions, it is natural to seek definitions which shall be independent of the use of any coördinates. It will be sufficient to give the following definitions without formal proof of their equivalence with those given above, since they are less convenient for use than those founded on systems of coördinates, and since we shall in fact have no occasion to use them.

We commence with the definition of extension-in-velocity. We may imagine n independent velocities, $V_1, \ldots V_n$ of which a system in a given configuration is capable. We may conceive of the system as having a certain velocity V_0 combined with a part of each of these velocities $V_1 \ldots V_n$. By a part of V_1 is meant a velocity of the same nature as V_1 but in amount being anything between zero and V_1. Now all the velocities which may be thus described may be regarded as forming or lying in a certain extension of which we desire a measure. The case is greatly simplified if we suppose that certain relations exist between the velocities $V_1, \ldots V_n$, viz: that the kinetic energy due to any two of these velocities combined is the sum of the kinetic energies due to the velocities separately. In this case the extension-in-motion is the square root of the product of the doubled kinetic energies due to the n velocities $V_1, \ldots V_n$ taken separately.

The more general case may be reduced to this simpler case as follows. The velocity V_2 may always be regarded as composed of two velocities V_2' and V_2'', of which V_2' is of the same nature as V_1, (it may be more or less in amount, or opposite in sign,) while V_2'' satisfies the relation that the kinetic energy due to V_1 and V_2'' combined is the sum of the kinetic energies due to these velocities taken separately. And the velocity V_3 may be regarded as compounded of three, V_3', V_3'', V_3''', of which V_3' is of the same nature as V_1, V_3'' of the same nature as V_2'', while V_3''' satisfies the relations that if combined either with V_1 or V_2'' the kinetic energy of the combined velocities is the sum of the kinetic energies of the velocities taken separately. When all the velocities $V_2, \ldots V_n$ have been thus decomposed, the square root of the product of the doubled kinetic energies of the several velocities V_1, V_2'', V_3''', etc., will be the value of the extension-in-velocity which is sought.

This method of evaluation of the extension-in-velocity which we are considering is perhaps the most simple and natural, but the result may be expressed in a more symmetrical form. Let us write ϵ_{12} for the kinetic energy of the velocities V_1 and V_2 combined, diminished by the sum of the kinetic energies due to the same velocities taken separately. This may be called the mutual energy of the velocities V_1 and V_2. Let the mutual energy of every pair of the velocities $V_1, \ldots V_n$ be expressed in the same way. Analogy would make ϵ_{11} represent the energy of twice V_1 diminished by twice the energy of V_1, *i. e.*, ϵ_{11} would represent twice the energy of V_1, although the term mutual energy is hardly appropriate to this case. At all events, let ϵ_{11} have this signification, and ϵ_{22} represent twice the energy of V_2, etc. The square root of the determinant

$$\begin{vmatrix} \epsilon_{11} & \epsilon_{12} & \ldots & \epsilon_{1n} \\ \epsilon_{21} & \epsilon_{22} & \ldots & \epsilon_{2n} \\ \cdot & \cdot & \cdot & \cdot \\ \epsilon_{n1} & \epsilon_{n2} & \ldots & \epsilon_{nn} \end{vmatrix}$$

represents the value of the extension-in-velocity determined as above described by the velocities $V_1, \ldots V_n$.

The statements of the preceding paragraph may be readily proved from the expression (157) on page 60, viz.,

$$\Delta_q^{\frac{1}{2}}\, d\dot{q}_1 \ldots d\dot{q}_n$$

by which the notion of an element of extension-in-velocity was originally defined. Since $\Delta_{\dot{q}}$ in this expression represents the determinant of which the general element is

$$\frac{d^2\epsilon}{d\dot{q}_i d\dot{q}_j}$$

the square of the preceding expression represents the determinant of which the general element is

$$\frac{d^2\epsilon}{d\dot{q}_i d\dot{q}_j} d\dot{q}_i d\dot{q}_j.$$

Now we may regard the differentials of velocity $d\dot{q}_i$, $d\dot{q}_j$ as themselves infinitesimal velocities. Then the last expression represents the mutual energy of these velocities, and

$$\frac{d^2\epsilon}{d\dot{q}_i^2} d\dot{q}_i^2$$

represents twice the energy due to the velocity $d\dot{q}_i$.

The case which we have considered is an extension-in-velocity of the simplest form. All extensions-in-velocity do not have this form, but all may be regarded as composed of elementary extensions of this form, in the same manner as all volumes may be regarded as composed of elementary parallelepipeds.

Having thus a measure of extension-in-velocity founded, it will be observed, on the dynamical notion of kinetic energy, and not involving an explicit mention of coördinates, we may derive from it a measure of extension-in-configuration by the principle connecting these quantities which has been given in a preceding paragraph of this chapter.

The measure of extension-in-phase may be obtained from that of extension-in-configuration and of extension-in-velocity. For to every configuration in an extension-in-phase there will belong a certain extension-in-velocity, and the integral of the elements of extension-in-configuration within any extension-in-phase multiplied each by its extension-in-velocity is the measure of the extension-in-phase.

CHAPTER VII.

FARTHER DISCUSSION OF AVERAGES IN A CANONICAL ENSEMBLE OF SYSTEMS.

RETURNING to the case of a canonical distribution, we have for the index of probability of configuration

$$\eta_q = \frac{\psi_q - \epsilon_q}{\Theta} \tag{178}$$

as appears on comparison of formulae (142) and (161). It follows immediately from (142) that the average value in the ensemble of any quantity u which depends on the configuration alone is given by the formula

$$\bar{u} = \int_{\text{config.}}^{\text{all}} \cdots \int u\, e^{\frac{\psi_q - \epsilon_q}{\Theta}} \Delta_q^{\frac{1}{2}} dq_1 \ldots dq_n, \tag{179}$$

where the integrations cover all possible configurations. The value of ψ_q is evidently determined by the equation

$$e^{-\frac{\psi_q}{\Theta}} = \int_{\text{config.}}^{\text{all}} \cdots \int e^{-\frac{\epsilon_q}{\Theta}} \Delta_q^{\frac{1}{2}} dq_1 \ldots dq_n. \tag{180}$$

By differentiating the last equation we may obtain results analogous to those obtained in Chapter IV from the equation

$$e^{-\frac{\psi}{\Theta}} = \int_{\text{phases}}^{\text{all}} \cdots \int e^{-\frac{\epsilon}{\Theta}} dp_1 \ldots dq_n.$$

As the process is identical, it is sufficient to give the results:

$$d\psi_q = \bar{\eta}_q d\Theta - \bar{A}_1 da_1 - \bar{A}_2 da_2 - \text{etc.}, \tag{181}$$

or, since
$$\psi_q = \bar{\epsilon}_q + \Theta \bar{\eta}_q, \tag{182}$$

and
$$d\psi_q = d\bar{\epsilon}_q + \bar{\eta}_q d\Theta + \Theta d\bar{\eta}_q, \tag{183}$$

$$d\bar{\epsilon}_q = -\Theta d\bar{\eta}_q - \bar{A}_1 da_1 - \bar{A}_2 da_2 - \text{etc.} \tag{184}$$

It appears from this equation that the differential relations subsisting between the average potential energy in an ensemble of systems canonically distributed, the modulus of distribution, the average index of probability of configuration, taken negatively, and the average forces exerted on external bodies, are equivalent to those enunciated by Clausius for the potential energy of a body, its temperature, a quantity which he called the disgregation, and the forces exerted on external bodies.*

For the index of probability of velocity, in the case of canonical distribution, we have by comparison of (144) and (163), or of (145) and (164),

$$\eta_p = \frac{\psi_p - \epsilon_p}{\Theta} \tag{185}$$

which gives
$$\bar{\eta}_p = \frac{\psi_p - \bar{\epsilon}_p}{\Theta}; \tag{186}$$

we have also
$$\bar{\epsilon}_p = \tfrac{1}{2} n \Theta, \tag{187}$$

and by (140),
$$\psi_p = -\tfrac{1}{2} n \Theta \log (2\pi\Theta). \tag{188}$$

From these equations we get by differentiation

$$d\psi_p = \bar{\eta}_p d\Theta, \tag{189}$$

and
$$d\bar{\epsilon}_p = -\Theta d\bar{\eta}_p. \tag{190}$$

The differential relation expressed in this equation between the average kinetic energy, the modulus, and the average index of probability of velocity, taken negatively, is identical with that given by Clausius *locis citatis* for the kinetic energy of a body, the temperature, and a quantity which he called the transformation-value of the kinetic energy.† The relations

$$\bar{\epsilon} = \bar{\epsilon}_q + \bar{\epsilon}_p, \qquad \bar{\eta} = \bar{\eta}_q + \bar{\eta}_p$$

* Pogg. Ann., Bd. CXVI, S. 73, (1862); ibid., Bd. CXXV, S. 353, (1865). See also Boltzmann, Sitzb. der Wiener Akad., Bd. LXIII, S. 728, (1871).

† Verwandlungswerth des Wärmeinhaltes.

are also identical with those given by Clausius for the corresponding quantities.

Equations (112) and (181) show that if ψ or ψ_q is known as function of Θ and a_1, a_2, etc., we can obtain by differentiation $\bar{\epsilon}$ or $\bar{\epsilon}_q$, and $\bar{A}_1$, $\bar{A}_2$, etc. as functions of the same variables. We have in fact

$$\bar{\epsilon} = \psi - \Theta \bar{\eta} = \psi - \Theta \frac{d\psi}{d\Theta}. \tag{191}$$

$$\bar{\epsilon}_q = \psi_q - \Theta \bar{\eta}_q = \psi_q - \Theta \frac{d\psi_q}{d\Theta}. \tag{192}$$

The corresponding equation relating to kinetic energy,

$$\bar{\epsilon}_p = \psi_p - \Theta \bar{\eta}_p = \psi_p - \Theta \frac{d\psi_p}{d\Theta}, \tag{193}$$

which may be obtained in the same way, may be verified by the known relations (186), (187), and (188) between the variables. We have also

$$\bar{A}_1 = -\frac{d\psi}{da_1} = -\frac{d\psi_q}{da_1}, \tag{194}$$

etc., so that the average values of the external forces may be derived alike from ψ or from ψ_q.

The average values of the squares or higher powers of the energies (total, potential, or kinetic) may easily be obtained by repeated differentiations of ψ, ψ_q, ψ_p, or $\bar{\epsilon}$, $\bar{\epsilon}_q$, $\bar{\epsilon}_p$, with respect to Θ. By equation (108) we have

$$\bar{\epsilon} = \int \underset{\text{phases}}{\overset{\text{all}}{\cdots}} \int \epsilon e^{\frac{\psi - \epsilon}{\Theta}} dp_1 \ldots dq_n, \tag{195}$$

and differentiating with respect to Θ,

$$\frac{d\bar{\epsilon}}{d\Theta} = \int \underset{\text{phases}}{\overset{\text{all}}{\cdots}} \int \left(\frac{\epsilon^2 - \psi\epsilon}{\Theta^2} + \frac{\epsilon}{\Theta}\frac{d\psi}{d\Theta} \right) e^{\frac{\psi - \epsilon}{\Theta}} dp_1 \ldots dq_n, \tag{196}$$

whence, again by (108),

$$\frac{d\bar{\epsilon}}{d\Theta} = \frac{\overline{\epsilon^2} - \psi\bar{\epsilon}}{\Theta^2} + \frac{\bar{\epsilon}}{\Theta}\frac{d\psi}{d\Theta},$$

or $$\overline{\epsilon^2} = \Theta^2 \frac{d\bar{\epsilon}}{d\Theta} + \bar{\epsilon}\left(\psi - \Theta \frac{d\psi}{d\Theta}\right). \tag{197}$$

Combining this with (191),

$$\overline{\epsilon^2} = \bar{\epsilon}^2 + \Theta^2 \frac{d\bar{\epsilon}}{d\Theta} = \left(\psi - \Theta\frac{d\psi}{d\Theta}\right)^2 - \Theta^3 \frac{d^2\psi}{d\Theta^2}. \tag{198}$$

In precisely the same way, from the equation

$$\bar{\epsilon}_q = \int_{\text{config.}}^{\text{all}} \cdots \int \epsilon_q e^{\frac{\psi_q - \epsilon_q}{\Theta}} \Delta_q^{\frac{1}{2}} dq_1 \ldots dq_n, \tag{199}$$

we may obtain

$$\overline{\epsilon_q^2} = \bar{\epsilon}_q^2 + \Theta^2 \frac{d\bar{\epsilon}_q}{d\Theta} = \left(\psi_q - \Theta\frac{d\psi_q}{d\Theta}\right)^2 - \Theta^3 \frac{d^2\psi_q}{d\Theta^2}. \tag{200}$$

In the same way also, if we confine ourselves to a particular configuration, from the equation

$$\bar{\epsilon}_p = \int_{\text{veloc.}}^{\text{all}} \cdots \int \epsilon_p e^{\frac{\psi_p - \epsilon_p}{\Theta}} \Delta_p^{\frac{1}{2}} dp_1 \ldots dp_n, \tag{201}$$

we obtain

$$\overline{\epsilon_p^2} = \bar{\epsilon}_p^2 + \Theta^2 \frac{d\bar{\epsilon}_p}{d\Theta} = \left(\psi_p - \Theta\frac{d\psi_p}{d\Theta}\right)^2 - \Theta^3 \frac{d^2\psi_p}{d\Theta^2}, \tag{202}$$

which by (187) reduces to

$$\overline{\epsilon_p^2} = (\tfrac{1}{4} n^2 + \tfrac{1}{2} n)\,\Theta^2. \tag{203}$$

Since this value is independent of the configuration, we see that the average square of the kinetic energy for every configuration is the same, and therefore the same as for the whole ensemble. Hence $\overline{\epsilon_p^2}$ may be interpreted as the average either for any particular configuration, or for the whole ensemble. It will be observed that the value of this quantity is determined entirely by the modulus and the number of degrees of freedom of the system, and is in other respects independent of the nature of the system.

Of especial importance are the anomalies of the energies, or their deviations from their average values. The average value of these anomalies is of course zero. The natural measure of such anomalies is the square root of their average square. Now

$$\overline{(\epsilon - \bar{\epsilon})^2} = \overline{\epsilon^2} - \bar{\epsilon}^2, \tag{204}$$

identically. Accordingly

$$\overline{(\epsilon - \bar{\epsilon})^2} = \Theta^2 \frac{d\bar{\epsilon}}{d\Theta} = - \Theta^3 \frac{d^2\psi}{d\Theta^2}. \tag{205}$$

In like manner,

$$\overline{(\epsilon_q - \bar{\epsilon}_q)^2} = \Theta^2 \frac{d\bar{\epsilon}_q}{d\Theta} = - \Theta^3 \frac{d^2\psi_q}{d\Theta^2}. \tag{206}$$

$$\overline{(\epsilon_p - \bar{\epsilon}_p)^2} = \Theta^2 \frac{d\bar{\epsilon}_p}{d\Theta} = - \Theta^3 \frac{d^2\psi_p}{d\Theta^2} = \tfrac{1}{2} n \, \Theta^2. \tag{207}$$

Hence

$$\overline{(\epsilon - \bar{\epsilon})^2} = \overline{(\epsilon_q - \bar{\epsilon}_q)^2} + \overline{(\epsilon_p - \bar{\epsilon}_p)^2}. \tag{208}$$

Equation (206) shows that the value of $d\bar{\epsilon}_q/d\Theta$ can never be negative, and that the value of $d^2\psi_q/d\Theta^2$ or $d\bar{\eta}_q/d\Theta$ can never be positive.*

To get an idea of the order of magnitude of these quantities, we may use the average kinetic energy as a term of comparison, this quantity being independent of the arbitrary constant involved in the definition of the potential energy. Since

* In the case discussed in the note on page 54, in which the potential energy is a quadratic function of the q's, and $\Delta_{\dot{q}}$ independent of the q's, we should get for the potential energy

$$\overline{(\epsilon_q - \bar{\epsilon}_q)^2} = \frac{1}{2} n \, \Theta^2,$$

and for the total energy

$$\overline{(\epsilon - \bar{\epsilon})^2} = n \, \Theta^2.$$

We may also write in this case,

$$\frac{\overline{(\epsilon_q - \bar{\epsilon}_q)^2}}{(\bar{\epsilon}_q - \epsilon_a)^2} = \frac{2}{n},$$

$$\frac{\overline{(\epsilon - \bar{\epsilon})^2}}{(\bar{\epsilon} - \epsilon_a)^2} = \frac{1}{n}.$$

$$\bar{\epsilon}_p = \tfrac{1}{2} n \Theta,$$

$$\frac{\overline{(\epsilon_p - \bar{\epsilon}_p)^2}}{\bar{\epsilon}_p^2} = \frac{2}{n}, \tag{209}$$

$$\frac{\overline{(\epsilon_q - \bar{\epsilon}_q)^2}}{\bar{\epsilon}_p^2} = \frac{2}{n} \frac{d\bar{\epsilon}_q}{d\bar{\epsilon}_p}, \tag{210}$$

$$\frac{\overline{(\epsilon - \bar{\epsilon})^2}}{\bar{\epsilon}_p^2} = \frac{2}{n} \frac{d\bar{\epsilon}}{d\bar{\epsilon}_p} = \frac{2}{n} + \frac{2}{n} \frac{d\bar{\epsilon}_q}{d\bar{\epsilon}_p}. \tag{211}$$

These equations show that when the number of degrees of freedom of the systems is very great, the mean squares of the anomalies of the energies (total, potential, and kinetic) are very small in comparison with the mean square of the kinetic energy, unless indeed the differential coefficient $d\bar{\epsilon}_q/d\bar{\epsilon}_p$ is of the same order of magnitude as n. Such values of $d\bar{\epsilon}_q/d\bar{\epsilon}_p$ can only occur within intervals $(\bar{\epsilon}_p'' - \bar{\epsilon}_p')$ which are of the order of magnitude of n^{-1}, unless it be in cases in which $\bar{\epsilon}_q$ is in general of an order of magnitude higher than $\bar{\epsilon}_p$. Postponing for the moment the consideration of such cases, it will be interesting to examine more closely the case of large values of $d\bar{\epsilon}_q/d\bar{\epsilon}_p$ within narrow limits. Let us suppose that for $\bar{\epsilon}_p'$ and $\bar{\epsilon}_p''$ the value of $d\bar{\epsilon}_q/d\bar{\epsilon}_p$ is of the order of magnitude of unity, but between these values of $\bar{\epsilon}_p$ very great values of the differential coefficient occur. Then in the ensemble having modulus Θ'' and average energies $\bar{\epsilon}_p''$ and $\bar{\epsilon}_q''$, values of ϵ_q sensibly greater than $\bar{\epsilon}_q''$ will be so rare that we may call them practically negligible. They will be still more rare in an ensemble of less modulus. For if we differentiate the equation

$$\eta_q = \frac{\psi_q - \epsilon_q}{\Theta}$$

regarding ϵ_q as constant, but Θ and therefore ψ_q as variable, we get

$$\left(\frac{d\eta_q}{d\Theta}\right)_{\epsilon_q} = \frac{1}{\Theta} \frac{d\psi_q}{d\Theta} - \frac{\psi_q - \epsilon_q}{\Theta^2}, \tag{212}$$

whence by (192)

$$\left(\frac{d\eta_q}{d\Theta}\right)_{\epsilon_q} = \frac{\epsilon_q - \bar{\epsilon}_q}{\Theta^2}. \tag{213}$$

That is, a diminution of the modulus will diminish the probability of all configurations for which the potential energy exceeds its average value in the ensemble. Again, in the ensemble having modulus Θ' and average energies $\bar{\epsilon}_p'$ and $\bar{\epsilon}_q'$, values of ϵ_q sensibly less than $\bar{\epsilon}_q'$ will be so rare as to be practically negligible. They will be still more rare in an ensemble of greater modulus, since by the same equation an increase of the modulus will diminish the probability of configurations for which the potential energy is less than its average value in the ensemble. Therefore, for values of Θ between Θ' and Θ'', and of $\bar{\epsilon}_p$ between $\bar{\epsilon}_p'$ and $\bar{\epsilon}_p''$, the individual values of ϵ_q will be practically limited to the interval between $\bar{\epsilon}_q'$ and $\bar{\epsilon}_q''$.

In the cases which remain to be considered, viz., when $d\bar{\epsilon}_q/d\bar{\epsilon}_p$ has very large values not confined to narrow limits, and consequently the differences of the mean potential energies in ensembles of different moduli are in general very large compared with the differences of the mean kinetic energies, it appears by (210) that the anomalies of mean square of potential energy, if not small in comparison with the mean kinetic energy, will yet in general be very small in comparison with differences of mean potential energy in ensembles having moderate differences of mean kinetic energy, — the exceptions being of the same character as described for the case when $d\bar{\epsilon}_q/d\bar{\epsilon}_p$ is not in general large.

It follows that to human experience and observation with respect to such an ensemble as we are considering, or with respect to systems which may be regarded as taken at random from such an ensemble, when the number of degrees of freedom is of such order of magnitude as the number of molecules in the bodies subject to our observation and experiment, $\epsilon - \bar{\epsilon}$, $\epsilon_p - \bar{\epsilon}_p$, $\epsilon_q - \bar{\epsilon}_q$ would be in general vanishing quantities, since such experience would not be wide enough to embrace the more considerable divergencies from the mean values, and such observation not nice enough to distinguish the ordinary divergencies. In other words, such ensembles would appear to human observation as ensembles of systems of uniform energy, and in which the potential and kinetic energies (supposing that there were means of measuring these quantities separately) had each separately uniform values.* Exceptions might occur when for particular values of the modulus the differential coefficient $d\bar{\epsilon}_q/d\bar{\epsilon}_p$ takes a very large value. To human observation the effect would be, that in ensembles in which Θ and $\bar{\epsilon}_p$ had certain critical values, $\bar{\epsilon}_q$ would be indeterminate within certain limits, viz., the values which would correspond to values of Θ and ϵ_p slightly less and slightly greater than the critical values. Such indeterminateness corresponds precisely to what we observe in experiments on the bodies which nature presents to us.†

To obtain general formulae for the average values of powers of the energies, we may proceed as follows. If h is any positive whole number, we have identically

$$\int \underset{\text{phases}}{\overset{\text{all}}{\cdots}} \int \epsilon^h e^{-\frac{\epsilon}{\Theta}} dp_1 \ldots dq_n = \Theta^2 \frac{d}{d\Theta} \int \underset{\text{phases}}{\overset{\text{all}}{\cdots}} \int \epsilon^{h-1} e^{-\frac{\epsilon}{\Theta}} dp_1 \ldots dq_n, \qquad (214)$$

i. e., by (108),

$$\overline{\epsilon^h} e^{-\frac{\psi}{\Theta}} = \Theta^2 \frac{d}{d\Theta} \left(\overline{\epsilon^{h-1}} e^{-\frac{\psi}{\Theta}} \right). \qquad (215)$$

Hence

$$\overline{\epsilon^h} e^{-\frac{\psi}{\Theta}} = \left(\Theta^2 \frac{d}{d\Theta} \right)^h e^{-\frac{\psi}{\Theta}}, \qquad (216)$$

and

$$\overline{\epsilon^h} = e^{\frac{\psi}{\Theta}} \left(\Theta^2 \frac{d}{d\Theta} \right)^h e^{-\frac{\psi}{\Theta}}. \qquad (217)$$

* This implies that the kinetic and potential energies of individual systems would each separately have values sensibly constant in time.

† As an example, we may take a system consisting of a fluid in a cylinder under a weighted piston, with a vacuum between the piston and the top of the cylinder, which is closed. The weighted piston is to be regarded as a part of the system. (This is formally necessary in order to satisfy the condition of the invariability of the external coördinates.) It is evident that at a certain temperature, viz., when the pressure of saturated vapor balances the weight of the piston, there is an indeterminateness in the values of the potential and total energies as functions of the temperature.

For $h = 1$, this gives

$$\bar{\epsilon} = -\Theta^2 \frac{d}{d\Theta}\left(\frac{\psi}{\Theta}\right) \tag{218}$$

which agrees with (191).
From (215) we have also

$$\overline{\epsilon^h} = \bar{\epsilon}\,\overline{\epsilon^{h-1}} + \Theta^2 \frac{d\overline{\epsilon^{h-1}}}{d\Theta} = \left(\bar{\epsilon} + \Theta^2 \frac{d}{d\Theta}\right)\overline{\epsilon^{h-1}}, \tag{219}$$

$$\overline{\epsilon^h} = \left(\bar{\epsilon} + \Theta^2 \frac{d}{d\Theta}\right)^{h-1} \bar{\epsilon}. \tag{220}$$

In like manner from the identical equation

$$\underset{\text{config.}}{\int \cdots \int^{\text{all}}} \epsilon_q^h e^{-\frac{\epsilon_q}{\Theta}} \Delta_q^{\frac{1}{2}} dq_1 \ldots dq_n = \Theta^2 \frac{d}{d\Theta} \underset{\text{config.}}{\int \cdots \int^{\text{all}}} \epsilon_q^{h-1} e^{-\frac{\epsilon_q}{\Theta}} \Delta_q^{\frac{1}{2}} dq_1 \ldots dq_n, \tag{221}$$

we get

$$\overline{\epsilon_q^h} = e^{\frac{\psi_q}{\Theta}} \left(\Theta^2 \frac{d}{d\Theta}\right)^h e^{-\frac{\psi_q}{\Theta}}, \tag{222}$$

and

$$\overline{\epsilon_q^h} = \left(\bar{\epsilon}_q + \Theta^2 \frac{d}{d\Theta}\right)^{h-1} \bar{\epsilon}_q. \tag{223}$$

With respect to the kinetic energy similar equations will hold for averages taken for any particular configuration, or for the whole ensemble. But since

$$\bar{\epsilon}_p = \frac{n}{2}\Theta,$$

the equation

$$\overline{\epsilon_p^h} = \left(\bar{\epsilon}_p + \Theta^2 \frac{d}{d\Theta}\right) \overline{\epsilon_p^{h-1}} \tag{224}$$

reduces to

$$\overline{\epsilon_p^h} = \left(\frac{n}{2}\Theta + {}^2\Theta \frac{d}{d\Theta}\right) \overline{\epsilon^{h-1}} = \frac{n}{2}\left(\frac{n}{2}\Theta + \Theta^2 \frac{d}{d\Theta}\right)^{h-1} \Theta. \tag{225}$$

We have therefore

$$\overline{\epsilon_p^2} = \left(\frac{n}{2}+1\right)\frac{n}{2}\,\Theta^2. \tag{226}$$

$$\overline{\epsilon_p^3} = \left(\frac{n}{2}+2\right)\left(\frac{n}{2}+1\right)\frac{n}{2}\,\Theta^3. \tag{227}$$

$$\overline{\epsilon_p^h} = \frac{\Gamma(\frac{1}{2}n+h)}{\Gamma(\frac{1}{2}n)}\,\Theta^h. \tag{*228}$$

The average values of the powers of the anomalies of the energies are perhaps most easily found as follows. We have identically, since $\bar{\epsilon}$ is a function of Θ, while ϵ is a function of the p's and q's,

$$\Theta^2\frac{d}{d\Theta}\int\limits_{\text{phases}}^{\text{all}}\cdots\int(\epsilon-\bar{\epsilon})^h e^{-\frac{\epsilon}{\Theta}}\,dp_1,\ldots dq_n =$$

$$\int\limits_{\text{phases}}^{\text{all}}\cdots\int\left[\epsilon(\epsilon-\bar{\epsilon})^h - h(\epsilon-\bar{\epsilon})^{h-1}\,\Theta^2\frac{d\bar{\epsilon}}{d\Theta}\right]e^{-\frac{\epsilon}{\Theta}}\,dp_1,\ldots dq_n \tag{229}$$

i. e., by (108),

$$\Theta^2\frac{d}{d\Theta}\left[\overline{(\epsilon-\bar{\epsilon})^h}\,e^{-\frac{\psi}{\Theta}}\right] = \left[\overline{\epsilon(\epsilon-\bar{\epsilon})^h} - h\,\overline{(\epsilon-\bar{\epsilon})^{h-1}}\,\Theta^2\frac{d\bar{\epsilon}}{d\Theta}\right]e^{-\frac{\psi}{\Theta}}, \tag{230}$$

* In the case discussed in the note on page 54 we may easily get

$$\overline{(\epsilon_q-\epsilon_a)^h} = \left(\bar{\epsilon}_q-\epsilon_a+\Theta^2\frac{d}{d\Theta}\right)\overline{(\epsilon_q-\epsilon_a)^{h-1}},$$

which, with

$$\bar{\epsilon}_q-\epsilon_a=\frac{n}{2}\Theta,$$

gives

$$\overline{(\epsilon_q-\epsilon_a)^h} = \left(\frac{n}{2}\Theta+\Theta^2\frac{d}{d\Theta}\right)\overline{(\epsilon_q-\epsilon_a)^{h-1}} = \frac{n}{2}\left(\frac{n}{2}\Theta+\Theta^2\frac{d}{d\Theta}\right)^{h-1}\Theta.$$

Hence

$$\overline{(\epsilon_q-\epsilon_a)^h} = \overline{\epsilon_p^h}.$$

Again

$$\overline{(\epsilon-\epsilon_a)^h} = \left(\bar{\epsilon}-\epsilon_a+\Theta^2\frac{d}{d\Theta}\right)\overline{(\epsilon-\epsilon_a)^{h-1}},$$

which with

$$\bar{\epsilon}-\epsilon_a = n\,\Theta$$

gives

$$\overline{(\epsilon-\epsilon_a)^h} = \left(n\,\Theta+\Theta^2\frac{d}{d\Theta}\right)\overline{(\epsilon-\epsilon_a)^{h-1}} = n\left(n\,\Theta+\Theta^2\frac{d}{d\Theta}\right)^{h-1}\Theta,$$

hence

$$\overline{(\epsilon-\epsilon_a)^h} = \frac{\Gamma(n+h)}{\Gamma(n)}\,\Theta^h.$$

or since by (218)

$$\Theta^2 \frac{d}{d\Theta} e^{-\frac{\psi}{\Theta}} = \bar{\epsilon}\, e^{-\frac{\psi}{\Theta}},$$

$$\Theta^2 \frac{d}{d\Theta} \overline{(\epsilon - \bar{\epsilon})^h} + \overline{(\epsilon - \bar{\epsilon})^h \bar{\epsilon}} = \overline{\epsilon\,(\epsilon - \bar{\epsilon})^h} - h\,\overline{(\epsilon - \bar{\epsilon})^{h-1}}\, \Theta^2 \frac{d\bar{\epsilon}}{d\Theta},$$

$$\overline{(\epsilon - \bar{\epsilon})^{h+1}} = \Theta^2 \frac{d}{d\Theta} \overline{(\epsilon - \bar{\epsilon})^h} + h\,\overline{(\epsilon - \bar{\epsilon})^{h-1}}\, \Theta^2 \frac{d\bar{\epsilon}}{d\Theta}. \qquad (231)$$

In precisely the same way we may obtain for the potential energy

$$\overline{(\epsilon_q - \bar{\epsilon}_q)^{h+1}} = \Theta^2 \frac{d}{d\Theta} \overline{(\epsilon_q - \bar{\epsilon}_q)^h} + h\,\overline{(\epsilon_q - \bar{\epsilon}_q)^{h-1}}\, \Theta^2 \frac{d\bar{\epsilon}_q}{d\Theta}. \qquad (232)$$

By successive applications of (231) we obtain

$$\overline{(\epsilon - \bar{\epsilon})^2} = D\bar{\epsilon}$$

$$\overline{(\epsilon - \bar{\epsilon})^3} = D^2\bar{\epsilon}$$

$$\overline{(\epsilon - \bar{\epsilon})^4} = D^3\bar{\epsilon} + 3\,(D\bar{\epsilon})^2$$

$$\overline{(\epsilon - \bar{\epsilon})^5} = D^4\bar{\epsilon} + 10\, D\bar{\epsilon}\, D^2\bar{\epsilon}$$

$$\overline{(\epsilon - \bar{\epsilon})^6} = D^5\bar{\epsilon} + 15\, D\bar{\epsilon}\, D^3\bar{\epsilon} + 10\,(D^2\bar{\epsilon})^2 + 15\,(D\bar{\epsilon})^3 \text{ etc.}$$

where D represents the operator $\Theta^2\, d/d\Theta$. Similar expressions relating to the potential energy may be derived from (232).

For the kinetic energy we may write similar equations in which the averages may be taken either for a single configuration or for the whole ensemble. But since

$$\frac{d\epsilon_p}{d\Theta} = \frac{n}{2}$$

the general formula reduces to

$$\overline{(\epsilon_p - \bar{\epsilon}_p)^{h+1}} = \Theta^2 \frac{d}{d\Theta} \overline{(\epsilon_p - \bar{\epsilon}_p)^h} + \tfrac{1}{2} n h \Theta^2\, \overline{(\epsilon_p - \bar{\epsilon}_p)^{h-1}} \qquad (233)$$

or

$$\frac{\overline{(\epsilon_p - \bar{\epsilon}_p)^{h+1}}}{\bar{\epsilon}_p^{\,h+1}} = \frac{2\Theta}{n} \frac{d}{d\Theta} \frac{\overline{(\epsilon_p - \bar{\epsilon}_p)^h}}{\bar{\epsilon}_p^{\,h}} + \frac{2h}{n} \frac{\overline{(\epsilon_p - \bar{\epsilon}_p)^h}}{\bar{\epsilon}_p^{\,h}} + \frac{2h}{n} \frac{\overline{(\epsilon_p - \bar{\epsilon}_p)^{h-1}}}{\bar{\epsilon}_p^{\,h-1}} \qquad (234)$$

But since identically

$$\frac{\overline{(\epsilon_p - \bar{\epsilon}_p)^0}}{\bar{\epsilon}_p{}^0} = 1, \qquad \frac{\overline{(\epsilon - \bar{\epsilon})^1}}{\bar{\epsilon}} = 0,$$

the value of the corresponding expression for any index will be independent of Θ and the formula reduces to

$$\overline{\left(\frac{\epsilon_p - \bar{\epsilon}_p}{\bar{\epsilon}_p}\right)^{h+1}} = \frac{2h}{n}\overline{\left(\frac{\epsilon_p - \bar{\epsilon}_p}{\bar{\epsilon}_p}\right)^{h}} + \frac{2h}{n}\overline{\left(\frac{\epsilon_p - \bar{\epsilon}_p}{\bar{\epsilon}_p}\right)^{h-1}} \tag{235}$$

we have therefore

$$\overline{\left(\frac{\epsilon_p - \bar{\epsilon}_p}{\bar{\epsilon}_p}\right)^{0}} = 1, \qquad \overline{\left(\frac{\epsilon_p - \bar{\epsilon}_p}{\bar{\epsilon}_p}\right)^{3}} = \frac{8}{n^2},$$

$$\overline{\left(\frac{\epsilon_p - \bar{\epsilon}_p}{\bar{\epsilon}_p}\right)^{1}} = 0, \qquad \overline{\left(\frac{\epsilon_p - \bar{\epsilon}_p}{\bar{\epsilon}_p}\right)^{4}} = \frac{48}{n^3} + \frac{12}{n^2},$$

$$\overline{\left(\frac{\epsilon_p - \bar{\epsilon}_p}{\bar{\epsilon}_p}\right)^{2}} = \frac{2}{n}, \qquad \text{etc.*}$$

It will be observed that when ψ or $\bar{\epsilon}$ is given as function of Θ, all averages of the form $\overline{\epsilon^h}$ or $\overline{(\epsilon - \bar{\epsilon})^h}$ are thereby deter-

* In the case discussed in the preceding foot-notes we get easily

$$\overline{(\epsilon_q - \bar{\epsilon}_q)^h} = \overline{(\epsilon_p - \bar{\epsilon}_p)^h},$$

and

$$\overline{\left(\frac{\epsilon_q - \bar{\epsilon}_q}{\bar{\epsilon}_q - \epsilon_a}\right)^h} = \overline{\left(\frac{\epsilon_p - \bar{\epsilon}_p}{\bar{\epsilon}_p}\right)^h}.$$

For the total energy we have in this case

$$\overline{\left(\frac{\epsilon - \bar{\epsilon}}{\bar{\epsilon} - \epsilon_a}\right)^{h+1}} = \frac{h}{n}\overline{\left(\frac{\epsilon - \bar{\epsilon}}{\bar{\epsilon} - \epsilon_a}\right)^{h}} + \frac{h}{n}\overline{\left(\frac{\epsilon - \bar{\epsilon}}{\bar{\epsilon} - \epsilon_a}\right)^{h-1}}.$$

$$\overline{\left(\frac{\epsilon - \bar{\epsilon}}{\bar{\epsilon} - \epsilon_a}\right)^{2}} = \frac{1}{n}. \qquad \overline{\left(\frac{\epsilon - \bar{\epsilon}}{\bar{\epsilon} - \epsilon_a}\right)^{4}} = \frac{3}{n^2} + \frac{6}{n^3},$$

$$\overline{\left(\frac{\epsilon - \bar{\epsilon}}{\bar{\epsilon} - \epsilon_a}\right)^{3}} = \frac{2}{n^2}, \qquad \text{etc.}$$

mined. So also if ψ_q or $\bar{\epsilon}_q$ is given as function of Θ, all averages of the form $\overline{\epsilon_q^h}$ or $\overline{(\epsilon_q - \bar{\epsilon}_q)^h}$ are determined. But

$$\bar{\epsilon}_q = \bar{\epsilon} - \tfrac{1}{2} n \Theta.$$

Therefore if any one of the quantities ψ, ψ_q, $\bar{\epsilon}$, $\bar{\epsilon}_q$ is known as function of Θ, and n is also known, all averages of any of the forms mentioned are thereby determined as functions of the same variable. In any case all averages of the form

$$\overline{\left(\frac{\epsilon_p - \bar{\epsilon}_p}{\bar{\epsilon}_p}\right)^h}$$

are known in terms of n alone, and have the same value whether taken for the whole ensemble or limited to any particular configuration.

If we differentiate the equation

$$\underset{\text{phases}}{\int \cdots \int^{\text{all}}} e^{\frac{\psi - \epsilon}{\Theta}} dp_1 \ldots dq_n = 1 \tag{236}$$

with respect to a_1, and multiply by Θ, we have

$$\int \cdots \int \left[\frac{d\psi}{da_1} - \frac{d\epsilon}{da_1}\right] e^{\frac{\psi - \epsilon}{\Theta}} dp_1 \ldots dq_n = 0. \tag{237}$$

Differentiating again, with respect to a_1, with respect to a_2, and with respect to Θ, we have

$$\int \cdots \int \left[\frac{d^2\psi}{da_1^{\,2}} - \frac{d^2\epsilon}{da_1^{\,2}} + \frac{1}{\Theta}\left(\frac{d\psi}{da_1} - \frac{d\epsilon}{da_1}\right)^2\right] e^{\frac{\psi - \epsilon}{\Theta}} dp_1 \ldots dq_n = 0, \tag{238}$$

$$\int \cdots \int \left[\frac{d^2\psi}{da_1\, da_2} - \frac{d^2\epsilon}{da_1\, da_2} + \frac{1}{\Theta}\left(\frac{d\psi}{da_1} - \frac{d\epsilon}{da_1}\right)\left(\frac{d\psi}{da_2} - \frac{d\epsilon}{da_2}\right)\right] e^{\frac{\psi - \epsilon}{\Theta}} dp_1 \ldots dq_n = 0, \tag{239}$$

$$\int \cdots \int \left[\frac{d^2\psi}{da_1\, d\Theta} + \left(\frac{d\psi}{da_1} - \frac{d\epsilon}{da_1}\right)\left(\frac{1}{\Theta}\frac{d\psi}{d\Theta} - \frac{\psi - \epsilon}{\Theta^2}\right)\right] e^{\frac{\psi - \epsilon}{\Theta}} dp_1 \ldots dq_n = 0. \tag{240}$$

The multiple integrals in the last four equations represent the average values of the expressions in the brackets, which we may therefore set equal to zero. The first gives

$$\frac{d\psi}{da_1} = \frac{\overline{d\epsilon}}{da_1} = -\bar{A}_1, \tag{241}$$

as already obtained. With this relation and (191) we get from the other equations

$$\overline{(A_1 - \bar{A}_1)^2} = \Theta\left(\frac{\overline{d^2\epsilon}}{da_1^2} - \frac{d^2\psi}{da_1^2}\right) = \Theta\left(\frac{d\bar{A}_1}{da_1} - \frac{\overline{dA_1}}{da_1}\right) \tag{242}$$

$$\begin{aligned} \overline{(A_1 - \bar{A}_1)(A_2 - \bar{A}_2)} &= \Theta\left(\frac{\overline{d^2\epsilon}}{da_1\,da_2} - \frac{d^2\psi}{da_1\,da_2}\right) \\ &= \Theta\left(\frac{d\bar{A}_1}{da_2} - \frac{\overline{dA_1}}{da_2}\right) = \Theta\left(\frac{d\bar{A}_2}{da_1} - \frac{\overline{dA_2}}{da_1}\right) \end{aligned} \tag{243}$$

$$\overline{(A_1 - \bar{A}_1)(\epsilon - \bar{\epsilon})} = -\Theta^2\frac{d^2\psi}{da_1\,d\Theta} = \Theta^2\frac{d\bar{A}_1}{d\Theta} = -\Theta^2\frac{d\bar{\eta}}{da_1}.$$

We may add for comparison equation (205), which might be derived from (236) by differentiating twice with respect to Θ:

$$\overline{(\epsilon - \bar{\epsilon})^2} = -\Theta^3\frac{d^2\psi}{d\Theta^2} = \Theta^2\frac{d\bar{\epsilon}}{d\Theta}. \tag{244}$$

The two last equations give

$$\overline{(A_1 - A_1)(\epsilon - \bar{\epsilon})} = \frac{d\bar{A}}{d\bar{\epsilon}}\overline{(\epsilon - \bar{\epsilon})^2}. \tag{245}$$

If ψ or $\bar{\epsilon}$ is known as function of Θ, a_1, a_2, etc., $\overline{(\epsilon - \bar{\epsilon})^2}$ may be obtained by differentiation as function of the same variables. And if ψ, or $\bar{A}_1$, or $\bar{\eta}$ is known as function of Θ, a_1, etc., $\overline{(A_1 - A_1)(\epsilon - \bar{\epsilon})}$ may be obtained by differentiation. But $\overline{(A_1 - \bar{A}_1)^2}$ and $\overline{(A_1 - \bar{A}_1)(A_2 - \bar{A}_2)}$ cannot be obtained in any similar manner. We have seen that $\overline{(\epsilon - \bar{\epsilon})^2}$ is in general a vanishing quantity for very great values of n, which we may regard as contained implicitly in Θ as a divisor. The same is true of $\overline{(A_1 - \bar{A}_1)(\epsilon - \bar{\epsilon})}$. It does not appear that we can assert the same of $\overline{(A_1 - \bar{A}_1)^2}$ or $\overline{(A_1 - \bar{A}_1)(A_2 - \bar{A}_2)}$, since

6

$d^2\epsilon/da_1^2$ may be very great. The quantities $d^2\epsilon/da_1^2$ and $d^2\psi/da_1^2$ belong to the class called elasticities. The former expression represents an elasticity measured under the condition that while a_1 is varied the internal coördinates $q_1, \ldots q_n$ all remain fixed. The latter is an elasticity measured under the condition that when a_1 is varied the ensemble remains canonically distributed within the same modulus. This corresponds to an elasticity in physics measured under the condition of constant temperature. It is evident that the former is greater than the latter, and it may be enormously greater.

The divergences of the force A_1 from its average value are due in part to the differences of energy in the systems of the ensemble, and in part to the differences in the value of the forces which exist in systems of the same energy. If we write $\overline{A_1}\rfloor_\epsilon$ for the average value of A_1 in systems of the ensemble which have any same energy, it will be determined by the equation

$$\overline{A_1}\rfloor_\epsilon = \frac{\int \cdots \int -\frac{d\epsilon}{da_1} e^{\frac{\psi-\epsilon}{\Theta}} dp_1 \ldots dq_n}{\int \cdots \int e^{\frac{\psi-\epsilon}{\Theta}} dp_1 \ldots dq_n} \tag{246}$$

where the limits of integration in both multiple integrals are two values of the energy which differ infinitely little, say ϵ and $\epsilon + d\epsilon$. This will make the factor $e^{\frac{\psi-\epsilon}{\Theta}}$ constant within the limits of integration, and it may be cancelled in the numerator and denominator, leaving

$$\overline{A_1}\rfloor_\epsilon = \frac{\int \cdots \int -\frac{d\epsilon}{da_1} dp_1 \ldots dq_n}{\int \cdots \int dp_1 \ldots dq_n} \tag{247}$$

where the integrals as before are to be taken between ϵ and $\epsilon + d\epsilon$. $\overline{A_1}\rfloor_\epsilon$ is therefore independent of Θ, being a function of the energy and the external coördinates.

Now we have identically

$$A_1 - \bar{A}_1 = (A_1 - \overline{A_1}|_\epsilon) + (\overline{A_1}|_\epsilon - \bar{A}_1),$$

where $A_1 - \overline{A_1}|_\epsilon$ denotes the excess of the force (tending to increase a_1) exerted by any system above the average of such forces for systems of the same energy. Accordingly,

$$\overline{(A_1 - \bar{A}_1)^2} = \overline{(A_1 - \overline{A_1}|_\epsilon)^2} + 2\,\overline{(A_1 - \overline{A_1}|_\epsilon)(\overline{A_1}|_\epsilon - \bar{A}_1)} + \overline{(\overline{A_1}|_\epsilon - \bar{A}_1)^2}.$$

But the average value of $(A_1 - \overline{A_1}|_\epsilon)(\overline{A_1}|_\epsilon - \bar{A}_1)$ for systems of the ensemble which have the same energy is zero, since for such systems the second factor is constant. Therefore the average for the whole ensemble is zero, and

$$\overline{(A_1 - \bar{A}_1)^2} = \overline{(A_1 - \overline{A_1}|_\epsilon)^2} + \overline{(\overline{A_1}|_\epsilon - \bar{A}_1)^2}. \qquad (248)$$

In the same way it may be shown that

$$\overline{(A_1 - \bar{A}_1)(\epsilon - \bar{\epsilon})} = \overline{(\overline{A_1}|_\epsilon - \bar{A}_1)(\epsilon - \bar{\epsilon})}. \qquad (249)$$

It is evident that in ensembles in which the anomalies of energy $\epsilon - \bar{\epsilon}$ may be regarded as insensible the same will be true of the quantities represented by $\overline{A_1}|_\epsilon - \bar{A}_1$.

The properties of quantities of the form $\overline{A_1}|_\epsilon$ will be farther considered in Chapter X, which will be devoted to ensembles of constant energy.

It may not be without interest to consider some general formulae relating to averages in a canonical ensemble, which embrace many of the results which have been given in this chapter.

Let u be any function of the internal and external coördinates with the momenta and modulus. We have by definition

$$\bar{u} = \int_{\text{phases}}^{\text{all}} \cdots \int u e^{\frac{\psi - \epsilon}{\Theta}} dp_1 \ldots dq_n \qquad (250)$$

If we differentiate with respect to Θ, we have

$$\frac{d\bar{u}}{d\Theta} = \int_{\text{phases}}^{\text{all}} \cdots \int \left(\frac{du}{d\Theta} - \frac{u}{\Theta^2}(\psi - \epsilon) + \frac{u}{\Theta}\frac{d\psi}{d\Theta} \right) e^{\frac{\psi - \epsilon}{\Theta}} dp_1 \ldots dq_n,$$

or
$$\frac{d\bar{u}}{d\Theta} = \overline{\frac{du}{d\Theta}} - \frac{\overline{u(\psi - \epsilon)}}{\Theta^2} + \frac{\bar{u}}{\Theta}\frac{d\psi}{d\Theta}. \tag{251}$$

Setting $u = 1$ in this equation, we get

$$\frac{d\psi}{d\Theta} = \frac{\psi - \bar{\epsilon}}{\Theta},$$

and substituting this value, we have

$$\frac{d\bar{u}}{d\Theta} = \overline{\frac{du}{d\Theta}} + \frac{\overline{u\epsilon}}{\Theta^2} - \frac{\bar{u}\,\bar{\epsilon}}{\Theta^2}$$

or
$$\Theta^2 \frac{d\bar{u}}{d\Theta} - \Theta^2 \overline{\frac{du}{d\Theta}} = \overline{u\epsilon} - \bar{u}\,\bar{\epsilon} = \overline{(u - \bar{u})\,(\epsilon - \bar{\epsilon})}. \tag{252}$$

If we differentiate equation (250) with respect to a (which may represent any of the external coördinates), and write A for the force $-\dfrac{d\epsilon}{dA}$, we get

$$\frac{d\bar{u}}{da} = \int \underset{\text{phases}}{\overset{\text{all}}{\ldots}} \int \left(\frac{du}{da} + \frac{u}{\Theta}\frac{d\psi}{da} + \frac{u}{\Theta}A\right) e^{\frac{\psi - \epsilon}{\Theta}}\, dp_1 \ldots dq_n$$

or
$$\frac{d\bar{u}}{da} = \overline{\frac{du}{da}} + \frac{\bar{u}}{\Theta}\frac{d\psi}{da} + \frac{\overline{uA}}{\Theta} \tag{253}$$

Setting $u = 1$ in this equation, we get

$$\frac{d\psi}{da} = -\bar{A}.$$

Substituting this value, we have

$$\frac{d\bar{u}}{da} = \overline{\frac{du}{da}} + \frac{\overline{uA}}{\Theta} - \frac{\bar{u}\bar{A}}{\Theta}, \tag{254}$$

or
$$\Theta\frac{d\bar{u}}{da} - \Theta\overline{\frac{du}{da}} = \overline{uA} - \bar{u}\,\bar{A} = \overline{(u - \bar{u})\,(A - \bar{A})}. \tag{255}$$

Repeated applications of the principles expressed by equations (252) and (255) are perhaps best made in the particular cases. Yet we may write (252) in this form

$$\overline{(\epsilon + D)(u - \bar{u})} = 0, \tag{256}$$

where D represents the operator $\Theta^2\, d/d\Theta$.

Hence

$$\overline{(\epsilon + D)^h (u - \bar{u})} = 0, \tag{257}$$

where h is any positive whole number. It will be observed, that since ϵ is not function of Θ, $(\epsilon + D)^h$ may be expanded by the binomial theorem. Or, we may write

$$\overline{(\epsilon + D)\, u} = (\bar{\epsilon} + D)\, \bar{u}, \tag{258}$$

whence

$$\overline{(\epsilon + D)^h\, u} = (\bar{\epsilon} + D)^h\, \bar{u}. \tag{259}$$

But the operator $(\bar{\epsilon} + D)^h$, although in some respects more simple than the operator without the average sign on the ϵ, cannot be expanded by the binomial theorem, since $\bar{\epsilon}$ is a function of Θ with the external coördinates.

So from equation (254) we have

$$\overline{\left(\frac{A}{\Theta} + \frac{d}{da}\right)(u - \bar{u})} = 0, \tag{260}$$

whence

$$\overline{\left(\frac{A}{\Theta} + \frac{d}{da}\right)^h (u - \bar{u})} = 0; \tag{261}$$

and

$$\overline{\left(\frac{A}{\Theta} + \frac{d}{da}\right) u} = \left(\frac{\bar{A}}{\Theta} + \frac{d}{da}\right) \bar{u}, \tag{262}$$

whence

$$\overline{\left(\frac{A}{\Theta} + \frac{d}{da}\right)^h u} = \left(\frac{\bar{A}}{\Theta} + \frac{d}{da}\right)^h \bar{u}. \tag{263}$$

The binomial theorem cannot be applied to these operators.

Again, if we now distinguish, as usual, the several external coördinates by suffixes, we may apply successively to the expression $u - \bar{u}$ any or all of the operators

$$\epsilon + \Theta^2 \frac{d}{d\Theta}, \quad A_1 + \Theta \frac{d}{da_1}, \quad A_2 + \Theta \frac{d}{da_2}, \quad \text{etc.} \tag{264}$$

as many times as we choose, and in any order, the average value of the result will be zero. Or, if we apply the same operators to u, and finally take the average value, it will be the same as the value obtained by writing the sign of average separately as u, and on ϵ, A_1, A_2, etc., in all the operators.

If u is independent of the momenta, formulae similar to the preceding, but having ϵ_q in place of ϵ, may be derived from equation (179).

CHAPTER VIII.

ON CERTAIN IMPORTANT FUNCTIONS OF THE ENERGIES OF A SYSTEM.

In order to consider more particularly the distribution of a canonical ensemble in energy, and for other purposes, it will be convenient to use the following definitions and notations.

Let us denote by V the extension-in-phase below a certain limit of energy which we shall call ϵ. That is, let

$$V = \int \dots \int dp_1 \dots dq_n, \tag{265}$$

the integration being extended (with constant values of the external coördinates) over all phases for which the energy is less than the limit ϵ. We shall suppose that the value of this integral is not infinite, except for an infinite value of the limiting energy. This will not exclude any kind of system to which the canonical distribution is applicable. For if

$$\int \dots \int e^{-\frac{\epsilon}{\Theta}} dp_1 \dots dq_n$$

taken without limits has a finite value,* the less value represented by

$$e^{-\frac{\epsilon}{\Theta}} \int \dots \int dp_1 \dots dq_n$$

taken below a limiting value of ϵ, and with the ϵ before the integral sign representing that limiting value, will also be finite. Therefore the value of V, which differs only by a constant factor, will also be finite, for finite ϵ. It is a function of ϵ and the external coördinates, a continuous increasing

* This is a necessary condition of the canonical distribution. See Chapter IV, p. 35.

function of ϵ, which becomes infinite with ϵ, and vanishes for the smallest possible value of ϵ, or for $\epsilon = -\infty$, if the energy may be diminished without limit.

Let us also set

$$\phi = \log \frac{dV}{d\epsilon}. \tag{266}$$

The extension in phase between any two limits of energy, ϵ' and ϵ'', will be represented by the integral

$$\int_{\epsilon'}^{\epsilon''} e^{\phi}\, d\epsilon. \tag{267}$$

And in general, we may substitute $e^{\phi}\, d\epsilon$ for $dp_1 \ldots dq_n$ in a $2n$-fold integral, reducing it to a simple integral, whenever the limits can be expressed by the energy alone, and the other factor under the integral sign is a function of the energy alone, or with quantities which are constant in the integration.

In particular we observe that the probability that the energy of an unspecified system of a canonical ensemble lies between the limits ϵ' and ϵ'' will be represented by the integral *

$$\int_{\epsilon'}^{\epsilon''} e^{\frac{\psi - \epsilon}{\Theta} + \phi}\, d\epsilon, \tag{268}$$

and that the average value in the ensemble of any quantity which only varies with the energy is given by the equation †

$$\bar{u} = \int_{V=0}^{\epsilon=\infty} u e^{\frac{\psi - \epsilon}{\Theta} + \phi}\, d\epsilon, \tag{269}$$

where we may regard the constant ψ as determined by the equation ‡

$$e^{-\frac{\psi}{\Theta}} = \int_{V=0}^{\epsilon=\infty} e^{-\frac{\epsilon}{\Theta} + \phi}\, d\epsilon, \tag{270}$$

In regard to the lower limit in these integrals, it will be observed that $V = 0$ is equivalent to the condition that the value of ϵ is the least possible.

* Compare equation (93). † Compare equation (108).

‡ Compare equation (92).

In like manner, let us denote by V_q the extension-in-configuration below a certain limit of potential energy which we may call ϵ_q. That is, let

$$V_q = \int \ldots \int \Delta_q^{\frac{1}{2}} \, dq_1 \ldots dq_n, \tag{271}$$

the integration being extended (with constant values of the external coōrdinates) over all configurations for which the potential energy is less than ϵ_q. V_q will be a function of ϵ_q with the external coördinates, an increasing function of ϵ_q, which does not become infinite (in such cases as we shall consider*) for any finite value of ϵ_q. It vanishes for the least possible value of ϵ_q, or for $\epsilon_q = -\infty$, if ϵ_q can be diminished without limit. It is not always a continuous function of ϵ_q. In fact, if there is a finite extension-in-configuration of constant potential energy, the corresponding value of V_q will not include that extension-in-configuration, but if ϵ_q be increased infinitesimally, the corresponding value of V_q will be increased by that finite extension-in-configuration.

Let us also set

$$\phi_q = \log \frac{dV_q}{d\epsilon_q}. \tag{272}$$

The extension-in-configuration between any two limits of potential energy ϵ_q' and ϵ_q'' may be represented by the integral

$$\int_{\epsilon_q'}^{\epsilon_q''} e^{\phi_q} \, d\epsilon_q \tag{273}$$

whenever there is no discontinuity in the value of V_q as function of ϵ_q between or at those limits, that is, whenever there is no finite extension-in-configuration of constant potential energy between or at the limits. And in general, with the restriction mentioned, we may substitute $e^{\phi_q} \, d\epsilon_q$ for $\Delta_q^{\frac{1}{2}} \, dq_1 \ldots dq_n$ in an n-fold integral, reducing it to a simple integral, when the limits are expressed by the potential energy, and the other factor under the integral sign is a function of

* If V_q were infinite for finite values of ϵ_q, V would evidently be infinite for finite values of ϵ.

the potential energy, either alone or with quantities which are constant in the integration.

We may often avoid the inconvenience occasioned by formulae becoming illusory on account of discontinuities in the values of V_q as function of ϵ_q by substituting for the given discontinuous function a continuous function which is practically equivalent to the given function for the purposes of the evaluations desired. It only requires infinitesimal changes of potential energy to destroy the finite extensions-in-configuration of constant potential energy which are the cause of the difficulty.

In the case of an ensemble of systems canonically distributed in configuration, when V_q is, or may be regarded as, a continuous function of ϵ_q (within the limits considered), the probability that the potential energy of an unspecified system lies between the limits ϵ_q' and ϵ_q'' is given by the integral

$$\int_{\epsilon_q'}^{\epsilon_q''} e^{\frac{\psi_q - \epsilon_q}{\Theta} + \phi_q} d\epsilon_q, \tag{274}$$

where ψ may be determined by the condition that the value of the integral is unity, when the limits include all possible values of ϵ_q. In the same case, the average value in the ensemble of any function of the potential energy is given by the equation

$$\bar{u} = \int_{V_q=0}^{\epsilon_q=\infty} u e^{\frac{\psi - \epsilon_q}{\Theta} + \phi_q} d\epsilon_q. \tag{275}$$

When V_q is not a continuous function of ϵ_q, we may write dV_q for $e^{\phi_q} d\epsilon_q$ in these formulae.

In like manner also, for any given configuration, let us denote by V_p the extension-in-velocity below a certain limit of kinetic energy specified by ϵ_p. That is, let

$$V_p = \int \dots \int \Delta_p^{\frac{1}{2}} \, dp_1 \dots dp_n, \tag{276}$$

the integration being extended, with constant values of the coördinates, both internal and external, over all values of the momenta for which the kinetic energy is less than the limit ϵ_p. V_p will evidently be a continuous increasing function of ϵ_p which vanishes and becomes infinite with ϵ_p. Let us set

$$\phi_p = \log \frac{dV_p}{d\epsilon_p}. \tag{277}$$

The extension-in-velocity between any two limits of kinetic energy ϵ_p' and ϵ_p'' may be represented by the integral

$$\int_{\epsilon_p'}^{\epsilon_p''} e^{\phi_p}\, d\epsilon_p. \tag{278}$$

And in general, we may substitute $e^{\phi_p}\, d\epsilon_p$ for $\Delta_p^{\frac{1}{2}}\, dp_1 \ldots dp_n$ or $\Delta_{\dot q}^{\frac{1}{2}}\, d\dot q_1 \ldots d\dot q_n$ in an n-fold integral in which the coördinates are constant, reducing it to a simple integral, when the limits are expressed by the kinetic energy, and the other factor under the integral sign is a function of the kinetic energy, either alone or with quantities which are constant in the integration.

It is easy to express V_p and ϕ_p in terms of ϵ_p. Since Δ_p is function of the coördinates alone, we have by definition

$$V_p = \Delta_p^{\frac{1}{2}} \int \ldots \int dp_1 \ldots dp_n \tag{279}$$

the limits of the integral being given by ϵ_p. That is, if

$$\epsilon_p = F(p_1, \ldots p_n), \tag{280}$$

the limits of the integral for $\epsilon_p = 1$, are given by the equation

$$F(p_1, \ldots p_n) = 1, \tag{281}$$

and the limits of the integral for $\epsilon_p = a^2$, are given by the equation

$$F(p_1, \ldots p_n) = a^2. \tag{282}$$

But since F represents a quadratic function, this equation may be written

$$F\left(\frac{p_1}{a}, \ldots \frac{p_n}{a}\right) = 1 \tag{283}$$

The value of V_p may also be put in the form

$$V_p = a^n \Delta_p^{\frac{1}{2}} \int \cdots \int d\frac{p_1}{a} \cdots d\frac{p_n}{a}. \tag{284}$$

Now we may determine V_p for $\epsilon_p = 1$ from (279) where the limits are expressed by (281), and V_p for $\epsilon_p = a^2$ from (284) taking the limits from (283). The two integrals thus determined are evidently identical, and we have

$$(V_p)_{\epsilon_p=a^2} = a^n (V_p)_{\epsilon_p=1} \tag{285}$$

i. e., V_p varies as $\epsilon_p^{\frac{n}{2}}$. We may therefore set

$$V_p = C\epsilon_p^{\frac{n}{2}}, \qquad e^{\phi_p} = \frac{n}{2} C \epsilon_p^{\frac{n}{2}-1} \tag{286}$$

where C is a constant, at least for fixed values of the internal coördinates.

To determine this constant, let us consider the case of a canonical distribution, for which we have

$$\int_0^\infty e^{\frac{\psi_p - \epsilon_p}{\Theta} + \phi_p} d\epsilon_p = 1,$$

where

$$e^{\frac{\psi_p}{\Theta}} = (2\pi\Theta)^{-\frac{n}{2}}.$$

Substituting this value, and that of e^{ϕ_p} from (286), we get

$$\frac{n}{2} C \int_0^\infty e^{-\frac{\epsilon_p}{\Theta}} \epsilon_p^{\frac{n}{2}-1} d\epsilon_p = (2\pi\Theta)^{\frac{n}{2}},$$

$$\frac{n}{2} C \int_0^\infty e^{-\frac{\epsilon_p}{\Theta}} \left(\frac{\epsilon_p}{\Theta}\right)^{\frac{n}{2}-1} d\left(\frac{\epsilon_p}{\Theta}\right) = (2\pi)^{\frac{n}{2}},$$

$$\frac{n}{2} C \Gamma\left(\frac{n}{2}\right) = (2\pi)^{\frac{n}{2}}, \tag{287}$$

$$C = \frac{(2\pi)^{\frac{n}{2}}}{\Gamma(\frac{1}{2}n + 1)}.$$

Having thus determined the value of the constant C, we may substitute it in the general expressions (286), and obtain the following values, which are perfectly general:

$$V_p = \frac{(2\pi\epsilon_p)^{\frac{n}{2}}}{\Gamma(\frac{1}{2}n+1)} \qquad (288)$$

$$e^{\phi_p} = \frac{(2\pi)^{\frac{n}{2}}\epsilon_p^{\frac{n}{2}-1}}{\Gamma(\frac{1}{2}n)} \qquad {}^*(289)$$

It will be observed that the values of V_p and ϕ_p for any given ϵ_p are independent of the configuration, and even of the nature of the system considered, except with respect to its number of degrees of freedom.

Returning to the canonical ensemble, we may express the probability that the kinetic energy of a system of a given configuration, but otherwise unspecified, falls within given limits, by either member of the following equation

$$\int e^{\frac{\psi_p-\epsilon_p}{\Theta}+\phi_p} d\epsilon_p = \frac{1}{\Gamma(\frac{1}{2}n)}\int e^{-\frac{\epsilon_p}{\Theta}}\left(\frac{\epsilon_p}{\Theta}\right)^{\frac{n}{2}-1} d\left(\frac{\epsilon_p}{\Theta}\right). \qquad (290)$$

Since this value is independent of the coördinates it also represents the probability that the kinetic energy of an unspecified system of a canonical ensemble falls within the limits. The form of the last integral also shows that the probability that the ratio of the kinetic energy to the modulus

* Very similar values for V_q, e^{ϕ_q}, V, and e^{ϕ} may be found in the same way in the case discussed in the preceding foot-notes (see pages 54, 72, 77, and 79), in which ϵ_q is a quadratic function of the q's, and $\Delta_{\dot q}$ independent of the q's. In this case we have

$$V_q = \left(\frac{\Delta_{\dot q}}{\Delta_q}\right)^{\frac{1}{2}} \frac{(2\pi)^{\frac{n}{2}}(\epsilon_q-\epsilon_a)^{\frac{n}{2}}}{\Gamma(\frac{1}{2}n+1)},$$

$$e^{\phi_q} = \left(\frac{\Delta_{\dot q}}{\Delta_q}\right)^{\frac{1}{2}} \frac{(2\pi)^{\frac{n}{2}}(\epsilon_q-\epsilon_a)^{\frac{n}{2}-1}}{\Gamma(\frac{1}{2}n)},$$

$$V = \left(\frac{\Delta_{\dot q}}{\Delta_q}\right)^{\frac{1}{2}} \frac{(2\pi)^{n}(\epsilon-\epsilon_a)^{n}}{\Gamma(n+1)},$$

$$e^{\phi} = \left(\frac{\Delta_{\dot q}}{\Delta_q}\right)^{\frac{1}{2}} \frac{(2\pi)^{n}(\epsilon-\epsilon_a)^{n-1}}{\Gamma(n)}.$$

falls within given limits is independent also of the value of the modulus, being determined entirely by the number of degrees of freedom of the system and the limiting values of the ratio.

The average value of any function of the kinetic energy, either for the whole ensemble, or for any particular configuration, is given by

$$\bar{u} = \frac{1}{\Theta^{\frac{n}{2}} \Gamma(\frac{1}{2}n)} \int_0^\infty u\, e^{-\frac{\epsilon_p}{\Theta}} \epsilon_p^{\frac{n}{2}-1}\, d\epsilon_p \qquad *(291)$$

Thus:

$$\overline{\epsilon_p{}^m} = \frac{\Gamma(m+\frac{1}{2}n)}{\Gamma(\frac{1}{2}n)} \Theta^m, \qquad \text{if} \qquad m + \tfrac{1}{2}n > 0; \qquad \dagger(292)$$

$$\overline{V_p} = \frac{\Gamma(n)}{\Gamma(\frac{1}{2}n+1)\,\Gamma(\frac{1}{2}n)} (2\pi\Theta)^{\frac{n}{2}}; \qquad (293)$$

* The corresponding equation for the average value of any function of the potential energy, when this is a quadratic function of the q's, and $\Delta_{\dot{q}}$ is independent of the q's, is

$$\bar{u} = \frac{1}{\Theta^{\frac{n}{2}} \Gamma(\frac{1}{2}n)} \int_{\epsilon_a}^\infty u\, e^{-\frac{\epsilon_q - \epsilon_a}{\Theta}} (\epsilon_q - \epsilon_a)^{\frac{n}{2}-1}\, d\epsilon_q.$$

In the same case, the average value of any function of the (total) energy is given by the equation

$$\bar{u} = \frac{1}{\Theta^n \Gamma(n)} \int_0^\infty u\, e^{-\frac{\epsilon - \epsilon_a}{\Theta}} (\epsilon - \epsilon_a)^{n-1}\, d\epsilon.$$

Hence in this case

$$\overline{(\epsilon_q - \epsilon_a)^m} = \frac{\Gamma(m+\frac{1}{2}n)}{\Gamma(\frac{1}{2}n)} \Theta^m, \qquad \text{if} \qquad m + \frac{1}{2}n > 0.$$

$$\overline{(\epsilon - \epsilon_a)^m} = \frac{\Gamma(m+n)}{\Gamma(n)} \Theta^m, \qquad \text{if} \qquad m + n > 0.$$

$$\overline{e^{-\phi_q} V_q} = \overline{e^{-\phi} V} = \Theta,$$

$$\overline{\frac{d\phi_q}{d\epsilon_q}} = \frac{1}{\Theta}, \qquad \text{if} \qquad n > 2,$$

and

$$\overline{\frac{d\phi}{d\epsilon}} = \frac{1}{\Theta}, \qquad \text{if} \qquad n > 1.$$

If $n = 1$, $e^\phi = 2\pi$ and $d\phi/d\epsilon = 0$ for any value of ϵ. If $n = 2$, the case is the same with respect to ϕ_q.

† This equation has already been proved for positive integral powers of the kinetic energy. See page 77.

$$\overline{e^{\phi_p}} = \frac{\Gamma(n-1)}{[\Gamma(\frac{1}{2}n)]^2} (2\pi)^{\frac{n}{2}} \Theta^{\frac{n}{2}-1}, \quad \text{if} \quad n > 1; \tag{294}$$

$$\overline{\frac{d\phi_p}{d\epsilon_p}} = \frac{1}{\Theta}, \quad \text{if} \quad n > 2; \tag{295}$$

$$\overline{e^{-\phi_p} V_p} = \Theta. \tag{296}$$

If $n = 2$, $e^{\phi_p} = 2\pi$, and $d\phi_p/d\epsilon_p = 0$, for any value of ϵ_p.

The definitions of V, V_q, and V_p give

$$V = \int\int d\,V_p\, d\,V_q \tag{297}$$

where the integrations cover all phases for which the energy is less than the limit ϵ, for which the value of V is sought. This gives

$$V = \int_{V_q=0}^{\epsilon_q=\epsilon} V_p\, d\,V_q, \tag{298}$$

and

$$e^{\phi} = \frac{d\,V}{d\epsilon} = \int_{V_q=0}^{\epsilon_q=\epsilon} e^{\phi_p} d\,V_q, \tag{299}$$

where V_p and e^{ϕ_p} are connected with V_q by the equation

$$\epsilon_p + \epsilon_q = \text{constant} = \epsilon. \tag{300}$$

If $n > 2$, e^{ϕ_p} vanishes at the upper limit, *i. e.*, for $\epsilon_p = 0$, and we get by another differentiation

$$e^{\phi} \frac{d\phi}{d\epsilon} = \int_{V_q=0}^{\epsilon_q=\epsilon} e^{\phi_p} \frac{d\phi_p}{d\epsilon_p} d\,V_q. \tag{301}$$

We may also write

$$V = \int_{V_q=0}^{\epsilon_q=\epsilon} V_p\, e^{\phi_q} d\epsilon_q, \tag{302}$$

$$e^{\phi} = \int_{V_q=0}^{\epsilon_q=\epsilon} e^{\phi_p+\phi_q} d\epsilon_q, \tag{303}$$

etc., when V_q is a continuous function of ϵ_q commencing with the value $V_q = 0$, or when we choose to attribute to V_q a fictitious continuity commencing with the value zero, as described on page 90.

If we substitute in these equations the values of V_p and e^{ϕ_p} which we have found, we get

$$V = \frac{(2\pi)^{\frac{n}{2}}}{\Gamma(\frac{1}{2}n+1)} \int_{V_q=0}^{\epsilon_q=\epsilon} (\epsilon - \epsilon_q)^{\frac{n}{2}} \, dV_q, \tag{304}$$

$$e^{\phi} = \frac{(2\pi)^{\frac{n}{2}}}{\Gamma(\frac{n}{2})} \int_{V_q=0}^{\epsilon_q=\epsilon} (\epsilon - \epsilon_q)^{\frac{n}{2}-1} \, dV_q, \tag{305}$$

where $e^{\phi_q} d\epsilon_q$ may be substituted for dV_q in the cases above described. If, therefore, n is known, and V_q as function of ϵ_q, V and e^{ϕ} may be found by quadratures.

It appears from these equations that V is always a continuous increasing function of ϵ, commencing with the value $V = 0$, even when this is not the case with respect to V_q and ϵ_q. The same is true of e^{ϕ}, when $n > 2$, or when $n = 2$ if V_q increases continuously with ϵ_q from the value $V_q = 0$.

The last equation may be derived from the preceding by differentiation with respect to ϵ. Successive differentiations give, if $h < \frac{1}{2}n + 1$,

$$\frac{d^h V}{d\epsilon^h} = \int_{V_q=0}^{\epsilon_q=\epsilon} \frac{d^h V_p}{d\epsilon_p{}^h} \, dV_q = \frac{(2\pi)^{\frac{n}{2}}}{\Gamma(\frac{1}{2}n+1-h)} \int_{V_q=0}^{\epsilon_q=\epsilon} (\epsilon - \epsilon_q)^{\frac{n}{2}-h} \, dV_q. \tag{306}$$

$d^h V / d\epsilon^h$ is therefore positive if $h < \frac{1}{2}n + 1$. It is an increasing function of ϵ, if $h < \frac{1}{2}n$. If ϵ is not capable of being diminished without limit, $d^h V / d\epsilon^h$ vanishes for the least possible value of ϵ, if $h < \frac{1}{2}n$.

If n is even,

$$\frac{d^{\frac{n}{2}} V}{d\epsilon^{\frac{n}{2}}} = (2\pi)^{\frac{n}{2}} (V_q)_{\epsilon_q=\epsilon} \tag{307}$$

That is, V_q is the same function of ϵ_q, as $\dfrac{1}{(2\pi)^{\frac{n}{2}}}\dfrac{d^{\frac{n}{2}}V}{d\epsilon^{\frac{n}{2}}}$ of ϵ.

When n is large, approximate formulae will be more available. It will be sufficient to indicate the method proposed, without precise discussion of the limits of its applicability or of the degree of its approximation. For the value of e^{ϕ} corresponding to any given ϵ, we have

$$e^{\phi} = \int_{V_q=0}^{\epsilon_q=\epsilon} e^{\phi_p+\phi_q}\, d\epsilon_q = \int_0^{\epsilon} e^{\phi_p+\phi_q}\, d\epsilon_p, \tag{308}$$

where the variables are connected by the equation (300). The maximum value of $\phi_p + \phi_q$ is therefore characterized by the equation

$$\frac{d\phi_p}{d\epsilon_p} = \frac{d\phi_q}{d\epsilon_q}. \tag{309}$$

The values of ϵ_p and ϵ_q determined by this maximum we shall distinguish by accents, and mark the corresponding values of functions of ϵ_p and ϵ_q in the same way. Now we have by Taylor's theorem

$$\phi_p = \phi_p' + \left(\frac{d\phi_p}{d\epsilon_p}\right)'(\epsilon_p - \epsilon_p') + \left(\frac{d^2\phi_p}{d\epsilon_p^{\,2}}\right)'\frac{(\epsilon_p - \epsilon_p')^2}{2} + \text{etc.} \tag{310}$$

$$\phi_q = \phi_q' + \left(\frac{d\phi_q}{d\epsilon_q}\right)'(\epsilon_q - \epsilon_q') + \left(\frac{d^2\phi_q}{d\epsilon_q^{\,2}}\right)'\frac{(\epsilon_q - \epsilon_q')}{2} + \text{etc.} \tag{311}$$

If the approximation is sufficient without going beyond the quadratic terms, since by (300)

$$\epsilon_p - \epsilon_p' = -(\epsilon_q - \epsilon_q'),$$

we may write

$$e^{\phi} = e^{\phi_p'+\phi_q'}\int_{-\infty}^{+\infty} e^{\left[\left(\frac{d^2\phi_p}{d\epsilon_p^{\,2}}\right)' + \left(\frac{d^2\phi_q}{d\epsilon_q^{\,2}}\right)'\right]\frac{(\epsilon_q-\epsilon_q')^2}{2}}\, d\epsilon_q, \tag{312}$$

where the limits have been made $\pm\,\infty$ for analytical simplicity. This is allowable when the quantity in the square brackets has a very large negative value, since the part of the integral corresponding to other than very small values of $\epsilon_q - \epsilon_q'$ may be regarded as a vanishing quantity.

This gives

$$e^{\phi} = e^{\phi_p' + \phi_q'} \left[\frac{-2\pi}{\left(\frac{d^2\phi_p}{d\epsilon_p^2}\right)' + \left(\frac{d^2\phi_q}{d\epsilon_q^2}\right)'} \right]^{\frac{1}{2}} \tag{313}$$

or

$$\phi = \phi_p' + \phi_q' + \tfrac{1}{2}\log(2\pi) - \tfrac{1}{2}\log\left[-\left(\frac{d^2\phi_p}{d\epsilon_p^2}\right)' - \left(\frac{d^2\phi_q}{d\epsilon_q^2}\right)' \right]. \tag{314}$$

From this equation, with (289), (300) and (309), we may determine the value of ϕ corresponding to any given value of ϵ, when ϕ_q is known as function of ϵ_q.

Any two systems may be regarded as together forming a third system. If we have V or ϕ given as function of ϵ for any two systems, we may express by quadratures V and ϕ for the system formed by combining the two. If we distinguish by the suffixes $(\ )_1$, $(\ )_2$, $(\ )_{12}$ the quantities relating to the three systems, we have easily from the definitions of these quantities

$$V_{12} = \int\int dV_1\, dV_2 = \int V_2\, dV_1 = \int V_1\, dV_2 = \int V_1 e^{\phi_2} d\epsilon_2, \tag{315}$$

$$e^{\phi_{12}} = \int e^{\phi_2} dV_1 = \int e^{\phi_1} dV_2 = \int e^{\phi_1 + \phi_2} d\epsilon_2, \tag{316}$$

where the double integral is to be taken within the limits

$$V_1 = 0,\ V_2 = 0, \text{ and } \epsilon_1 + \epsilon_2 = \epsilon_{12},$$

and the variables in the single integrals are connected by the last of these equations, while the limits are given by the first two, which characterize the least possible values of ϵ_1 and ϵ_2 respectively.

It will be observed that these equations are identical in form with those by which V and ϕ are derived from V_p or ϕ_p and V_q or ϕ_q, except that they do not admit in the general case those transformations which result from substituting for V_p or ϕ_p the particular functions which these symbols always represent.

Similar formulae may be used to derive V_q or ϕ_q for the compound system, when one of these quantities is known as function of the potential energy in each of the systems combined.

The operation represented by such an equation as

$$e^{\phi_{12}} = \int e^{\phi_1} e^{\phi_2} d\epsilon_1$$

is identical with one of the fundamental operations of the theory of errors, viz., that of finding the probability of an error from the probabilities of partial errors of which it is made up. It admits a simple geometrical illustration.

We may take a horizontal line as an axis of abscissas, and lay off ϵ_1 as an abscissa measured to the right of any origin, and erect e^{ϕ_1} as a corresponding ordinate, thus determining a certain curve. Again, taking a different origin, we may lay off ϵ_2 as abscissas measured to the left, and determine a second curve by erecting the ordinates e^{ϕ_2}. We may suppose the distance between the origins to be ϵ_{12}, the second origin being to the right if ϵ_{12} is positive. We may determine a third curve by erecting at every point in the line (between the least values of ϵ_1 and ϵ_2) an ordinate which represents the product of the two ordinates belonging to the curves already described. The area between this third curve and the axis of abscissas will represent the value of $e^{\phi_{12}}$. To get the value of this quantity for varying values of ϵ_{12}, we may suppose the first two curves to be rigidly constructed, and to be capable of being moved independently. We may increase or diminish ϵ_{12} by moving one of these curves to the right or left. The third curve must be constructed anew for each different value of ϵ_{12}.

CHAPTER IX.

THE FUNCTION ϕ AND THE CANONICAL DISTRIBUTION.

IN this chapter we shall return to the consideration of the canonical distribution, in order to investigate those properties which are especially related to the function of the energy which we have denoted by ϕ.

If we denote by N, as usual, the total number of systems in the ensemble,

$$N e^{\frac{\psi-\epsilon}{\Theta}+\phi} d\epsilon$$

will represent the number having energies between the limits ϵ and $\epsilon + d\epsilon$. The expression

$$N e^{\frac{\psi-\epsilon}{\Theta}+\phi} \qquad (317)$$

represents what may be called the *density-in-energy*. This vanishes for $\epsilon = \infty$, for otherwise the necessary equation

$$\int_{V=0}^{\epsilon=\infty} e^{\frac{\psi-\epsilon}{\Theta}+\phi} d\epsilon = 1 \qquad (318)$$

could not be fulfilled. For the same reason the density-in-energy will vanish for $\epsilon = -\infty$, if that is a possible value of the energy. Generally, however, the least possible value of the energy will be a finite value, for which, if $n > 2$, e^{ϕ} will vanish,* and therefore the density-in-energy. Now the density-in-energy is necessarily positive, and since it vanishes for extreme values of the energy if $n > 2$, it must have a maximum in such cases, in which the energy may be said to have

* See page 96.

its most common or most probable value, and which is determined by the equation

$$\frac{d\phi}{d\epsilon} = \frac{1}{\Theta}. \tag{319}$$

This value of $d\phi/d\epsilon$ is also, when $n > 2$, its average value in the ensemble. For we have identically, by integration by parts,

$$\int_{V=0}^{\epsilon=\infty} \frac{d\phi}{d\epsilon} e^{\frac{\psi-\epsilon}{\Theta}+\phi} d\epsilon = \left[e^{\frac{\psi-\epsilon}{\Theta}+\phi} \right]_{V=0}^{\epsilon=\infty} + \frac{1}{\Theta} \int_{V=0}^{\epsilon=\infty} e^{\frac{\psi-\epsilon}{\Theta}+\phi} d\epsilon. \tag{320}$$

If $n > 2$, the expression in the brackets, which multiplied by N would represent the density-in-energy, vanishes at the limits, and we have by (269) and (318)

$$\overline{\frac{d\phi}{d\epsilon}} = \frac{1}{\Theta}. \tag{321}$$

It appears, therefore, that for systems of more than two degrees of freedom, the average value of $d\phi/d\epsilon$ in an ensemble canonically distributed is identical with the value of the same differential coefficient as calculated for the most common energy in the ensemble, both values being reciprocals of the modulus.

Hitherto, in our consideration of the quantities V, V_q, V_p, ϕ, ϕ_q, ϕ_p, we have regarded the external coördinates as constant. It is evident, however, from their definitions that V and ϕ are in general functions of the external coördinates and the energy (ϵ), that V_q and ϕ_q are in general functions of the external coördinates and the potential energy (ϵ_q). V_p and ϕ_p we have found to be functions of the kinetic energy (ϵ_p) alone. In the equation

$$e^{-\frac{\psi}{\Theta}} = \int_{V=0}^{\epsilon=\infty} e^{-\frac{\epsilon}{\Theta}+\phi} d\epsilon, \tag{322}$$

by which ψ may be determined, Θ and the external coördinates (contained implicitly in ϕ) are constant in the integration. The equation shows that ψ is a function of these constants.

If their values are varied, we shall have by differentiation, if $n > 2$,

$$e^{-\frac{\psi}{\Theta}}\left(-\frac{1}{\Theta}\,d\psi + \frac{\psi}{\Theta^2}\,d\Theta\right) = \frac{1}{\Theta^2}\,d\Theta\int_{V=0}^{\epsilon=\infty} \epsilon\, e^{-\frac{\epsilon}{\Theta}+\phi}\,d\epsilon$$

$$+ da_1\int_{V=0}^{\epsilon=\infty}\frac{d\phi}{da_1}\,e^{-\frac{\epsilon}{\Theta}+\phi}\,d\epsilon + da_2\int_{V=0}^{\epsilon=\infty}\frac{d\phi}{da_2}\,e^{-\frac{\epsilon}{\Theta}+\phi}\,d\epsilon + \text{etc.} \quad (323)$$

(Since e^{ϕ} vanishes with V, when $n > 2$, there are no terms due to the variations of the limits.) Hence by (269)

$$-\frac{1}{\Theta}\,d\psi + \frac{\psi}{\Theta^2}\,d\Theta = \frac{\overline{\epsilon}}{\Theta^2}\,d\Theta + \overline{\frac{d\phi}{da_1}}\,da_1 + \overline{\frac{d\phi}{da_2}}\,da_2 + \text{etc.}, \quad (324)$$

or, since

$$\frac{\psi + \overline{\epsilon}}{\Theta} = \overline{\eta}, \quad (325)$$

$$d\psi = \overline{\eta}\,d\Theta - \Theta\,\overline{\frac{d\phi}{da_1}}\,da_1 - \Theta\,\overline{\frac{d\phi}{da_2}}\,da_2 - \text{etc.} \quad (326)$$

Comparing this with (112), we get

$$\overline{\frac{d\phi}{da_1}} = \frac{\overline{A}_1}{\Theta}, \quad \overline{\frac{d\phi}{da_2}} = \frac{\overline{A}_2}{\Theta}, \quad \text{etc.} \quad (327)$$

The first of these equations might be written*

$$\overline{\left(\frac{d\phi}{da_1}\right)_{\epsilon,a}} = -\overline{\left(\frac{d\phi}{d\epsilon}\right)_a}\,\overline{\left(\frac{d\epsilon}{da_1}\right)_{a,q}} \quad (328)$$

but must not be confounded with the equation

$$\overline{\left(\frac{d\phi}{da_1}\right)_{\epsilon,a}} = -\overline{\left(\frac{d\phi}{d\epsilon}\right)_a\left(\frac{d\epsilon}{da_1}\right)_{\phi,a}} \quad (329)$$

which is derived immediately from the identity

$$\left(\frac{d\phi}{da_1}\right)_{\epsilon,a} = -\left(\frac{d\phi}{d\epsilon}\right)_a\left(\frac{d\epsilon}{da_1}\right)_{\phi,a} \quad (330)$$

* See equations (321) and (104). Suffixes are here added to the differential coefficients, to make the meaning perfectly distinct, although the same quantities may be written elsewhere without the suffixes, when it is believed that there is no danger of misapprehension. The suffixes indicate the quantities which are constant in the differentiation, the single letter a standing for all the letters a_1, a_2, etc., or all except the one which is explicitly varied.

Moreover, if we eliminate $d\psi$ from (326) by the equation

$$d\psi = \Theta d\bar{\eta} + \bar{\eta} d\Theta + d\bar{\epsilon}, \tag{331}$$

obtained by differentiating (325), we get

$$d\bar{\epsilon} = - \Theta d\bar{\eta} - \Theta \overline{\frac{d\phi}{da_1}} da_1 - \Theta \overline{\frac{d\phi}{da_2}} da_2 - \text{etc.}, \tag{332}$$

or by (321),

$$- d\bar{\eta} = \overline{\frac{d\phi}{d\epsilon}} d\bar{\epsilon} + \overline{\frac{d\phi}{da_1}} da_1 + \overline{\frac{d\phi}{da_2}} da_2 + \text{etc.} \tag{333}$$

Except for the signs of average, the second member of this equation is the same as that of the identity

$$d\phi = \frac{d\phi}{d\epsilon} d\epsilon + \frac{d\phi}{da_1} da_1 + \frac{d\phi}{da_2} da_2 + \text{etc.} \tag{334}$$

For the more precise comparison of these equations, we may suppose that the energy in the last equation is some definite and fairly representative energy in the ensemble. For this purpose we might choose the average energy. It will perhaps be more convenient to choose the most common energy, which we shall denote by ϵ_0. The same suffix will be applied to functions of the energy determined for this value. Our identity then becomes

$$d\phi_0 = \left(\frac{d\phi}{d\epsilon}\right)_0 d\epsilon_0 + \left(\frac{d\phi}{da_1}\right)_0 da_1 + \left(\frac{d\phi}{da_2}\right)_0 da_2 + \text{etc.} \tag{335}$$

It has been shown that

$$\overline{\frac{d\phi}{d\epsilon}} = \left(\frac{d\phi}{d\epsilon}\right)_0 = \frac{1}{\Theta}, \tag{336}$$

when $n > 2$. Moreover, since the external coördinates have constant values throughout the ensemble, the values of $d\phi/da_1$, $d\phi/da_2$, etc. vary in the ensemble only on account of the variations of the energy (ϵ), which, as we have seen, may be regarded as sensibly constant throughout the ensemble, when n is very great. In this case, therefore, we may regard the average values

$$\overline{\frac{d\phi}{da_1}}, \quad \overline{\frac{d\phi}{da_2}}, \quad \text{etc.,}$$

as practically equivalent to the values relating to the most common energy

$$\left(\frac{d\phi}{da_1}\right)_0, \quad \left(\frac{d\phi}{da_2}\right)_0, \quad \text{etc.}$$

In this case also $d\bar{\epsilon}$ is practically equivalent to $d\epsilon_0$. We have therefore, for very large values of n,

$$-d\bar{\eta} = d\phi_0 \tag{337}$$

approximately. That is, except for an additive constant, $-\bar{\eta}$ may be regarded as practically equivalent to ϕ_0, when the number of degrees of freedom of the system is very great. It is not meant by this that the variable part of $\bar{\eta} + \phi_0$ is numerically of a lower order of magnitude than unity. For when n is very great, $-\bar{\eta}$ and ϕ_0 are very great, and we can only conclude that the variable part of $\bar{\eta} + \phi_0$ is insignificant compared with the variable part of $\bar{\eta}$ or of ϕ_0, taken separately.

Now we have already noticed a certain correspondence between the quantities Θ and $\bar{\eta}$ and those which in thermodynamics are called temperature and entropy. The property just demonstrated, with those expressed by equation (336), therefore suggests that the quantities ϕ and $d\epsilon/d\phi$ may also correspond to the thermodynamic notions of entropy and temperature. We leave the discussion of this point to a subsequent chapter, and only mention it here to justify the somewhat detailed investigation of the relations of these quantities.

We may get a clearer view of the limiting form of the relations when the number of degrees of freedom is indefinitely increased, if we expand the function ϕ in a series arranged according to ascending powers of $\epsilon - \epsilon_0$. This expansion may be written

$$\phi = \phi_0 + \left(\frac{d\phi}{d\epsilon}\right)_0 (\epsilon - \epsilon_0) + \left(\frac{d^2\phi}{d\epsilon^2}\right)_0 \frac{(\epsilon - \epsilon_0)^2}{2} + \left(\frac{d^3\phi}{d\epsilon^3}\right)_0 \frac{(\epsilon - \epsilon_0)^3}{\underline{|3}} + \text{etc.} \tag{338}$$

Adding the identical equation

$$\frac{\psi - \epsilon}{\Theta} = \frac{\psi - \epsilon_0}{\Theta} - \frac{\epsilon - \epsilon_0}{\Theta},$$

we get by (336)

$$\frac{\psi - \epsilon}{\Theta} + \phi = \frac{\psi - \epsilon_0}{\Theta} + \phi_0 + \left(\frac{d^2\phi}{d\epsilon^2}\right) \frac{(\epsilon - \epsilon_0)^2}{2} + \left(\frac{d^3\phi}{d\epsilon^3}\right) \frac{(\epsilon - \epsilon_0)^3}{\underline{|3}} + \text{etc.} \tag{339}$$

Substituting this value in

$$\int_{\epsilon'}^{\epsilon''} e^{\frac{\psi - \epsilon}{\Theta} + \phi} d\epsilon,$$

which expresses the probability that the energy of an unspecified system of the ensemble lies between the limits ϵ' and ϵ'', we get

$$e^{\frac{\psi - \epsilon_0}{\Theta} + \phi_0} \int_{\epsilon'}^{\epsilon''} e^{\left(\frac{d^2\phi}{d\epsilon^2}\right)_0 \frac{(\epsilon - \epsilon_0)^2}{2} + \left(\frac{d^3\phi}{d\epsilon^3}\right)_0 \frac{(\epsilon - \epsilon_0)^3}{\underline{|3}} + \text{etc.}} d\epsilon. \tag{340}$$

When the number of degrees of freedom is very great, and $\epsilon - \epsilon_0$ in consequence very small, we may neglect the higher powers and write*

$$e^{\frac{\psi - \epsilon_0}{\Theta} + \phi_0} \int_{\epsilon'}^{\epsilon''} e^{\left(\frac{d^2\phi}{d\epsilon^2}\right)_0 \frac{(\epsilon - \epsilon_0)^2}{2}} d\epsilon. \tag{341}$$

This shows that for a very great number of degrees of freedom the probability of deviations of energy from the most probable value (ϵ_0) approaches the form expressed by the 'law of errors.' With this approximate law, we get

* If a higher degree of accuracy is desired than is afforded by this formula, it may be multiplied by the series obtained from

$$e^{\left(\frac{d^3\phi}{d\epsilon^3}\right)_0 \frac{(\epsilon - \epsilon_0)^3}{\underline{|3}} + \text{etc.}}$$

by the ordinary formula for the expansion in series of an exponential function. There would be no especial analytical difficulty in taking account of a moderate number of terms of such a series, which would commence

$$1 + \left(\frac{d^3\phi}{d\epsilon^3}\right)_0 \frac{(\epsilon - \epsilon_0)^3}{\underline{|3}} + \left(\frac{d^4\phi}{d\epsilon^4}\right)_0 \frac{(\epsilon - \epsilon_0)^4}{\underline{|4}} + \text{ etc.}$$

$$e^{\frac{\psi-\epsilon_0}{\Theta}+\phi_0}\left(\frac{-2\pi}{\left(\frac{d^2\phi}{d\epsilon^2}\right)_0}\right)^{\frac{1}{2}}=1, \tag{342}$$

$$\bar{\epsilon}=\epsilon_0, \qquad \overline{(\epsilon-\epsilon_0)^2}=-\left(\frac{d^2\phi}{d\epsilon^2}\right)^{-1}, \tag{343}$$

whence

$$\frac{\psi-\epsilon_0}{\Theta}+\phi_0=\tfrac{1}{2}\log\frac{\left(\frac{d^2\phi}{d\epsilon^2}\right)_0}{-2\pi}=-\tfrac{1}{2}\log\left(2\pi\overline{(\epsilon-\bar{\epsilon})^2}\right). \tag{344}$$

Now it has been proved in Chapter VII that

$$\overline{(\epsilon-\bar{\epsilon})^2}=\frac{2}{n}\frac{d\bar{\epsilon}}{d\bar{\epsilon}_p}\bar{\epsilon}_p^{\,2}.$$

We have therefore

$$\bar{\eta}+\phi_0=\frac{\psi-\bar{\epsilon}}{\Theta}+\phi_0=-\tfrac{1}{2}\log\left(2\pi\overline{(\epsilon-\bar{\epsilon})^2}\right)=-\tfrac{1}{2}\log\left(\frac{4\pi}{n}\frac{d\bar{\epsilon}}{d\bar{\epsilon}_p}\bar{\epsilon}_p^{\,2}\right) \tag{345}$$

approximately. The order of magnitude of $\bar{\eta}-\phi_0$ is therefore that of $\log n$. This magnitude is mainly constant. The order of magnitude of $\bar{\eta}+\phi_0-\frac{1}{2}\log n$ is that of unity. The order of magnitude of ϕ_0, and therefore of $-\bar{\eta}$, is that of n.*

Equation (338) gives for the first approximation

$$\bar{\phi}-\phi_0=\left(\frac{d^2\phi}{d\epsilon^2}\right)_0\frac{\overline{(\epsilon-\epsilon_0)^2}}{2}=-\tfrac{1}{2}, \tag{346}$$

$$\overline{(\phi-\phi_0)(\epsilon-\epsilon_0)}=\frac{\overline{(\epsilon-\epsilon_0)^2}}{\Theta}=\frac{d\bar{\epsilon}}{d\bar{\epsilon}_p}\bar{\epsilon}_p, \tag{347}$$

$$\overline{(\phi-\phi_0)^2}=\frac{\overline{(\epsilon-\epsilon_0)^2}}{\Theta^2}=\frac{n}{2}\frac{d\bar{\epsilon}}{d\bar{\epsilon}_p} \tag{348}$$

The members of the last equation have the order of magnitude of n. Equation (338) gives also for the first approximation

$$\frac{d\phi}{d\epsilon}-\frac{1}{\Theta}=\left(\frac{d^2\phi}{d\epsilon^2}\right)_0(\epsilon-\epsilon_0),$$

* Compare (289), (314).

whence

$$\overline{\left(\frac{d\phi}{d\epsilon} - \frac{1}{\Theta}\right)(\epsilon - \epsilon_0)} = \left(\frac{d^2\phi}{d\epsilon^2}\right)_0 \overline{(\epsilon - \epsilon_0)^2} = -1, \tag{349}$$

$$\overline{\left(\frac{d\phi}{d\epsilon} - \frac{1}{\Theta}\right)^2} = \left(\frac{d^2\phi}{d\epsilon^2}\right)_0^2 \overline{(\epsilon - \epsilon_0)^2} = \frac{1}{\overline{(\epsilon - \epsilon_0)^2}} = -\left(\frac{d^2\phi}{d\epsilon^2}\right)_0. \tag{350}$$

This is of the order of magnitude of n.*

It should be observed that the approximate distribution of the ensemble in energy according to the 'law of errors' is not dependent on the particular form of the function of the energy which we have assumed for the index of probability (η). In any case, we must have

$$\int_{V=0}^{\epsilon=\infty} e^{\eta+\phi}\, d\epsilon = 1, \tag{351}$$

where $e^{\eta+\phi}$ is necessarily positive. This requires that it shall vanish for $\epsilon = \infty$, and also for $\epsilon = -\infty$, if this is a possible value. It has been shown in the last chapter that if ϵ has a (finite) least possible value (which is the usual case) and $n > 2$, e^{ϕ} will vanish for that least value of ϵ. In general therefore $\eta + \phi$ will have a maximum, which determines the most probable value of the energy. If we denote this value by ϵ_0, and distinguish the corresponding values of the functions of the energy by the same suffix, we shall have

$$\left(\frac{d\eta}{d\epsilon}\right)_0 + \left(\frac{d\phi}{d\epsilon}\right)_0 = 0. \tag{352}$$

The probability that an unspecified system of the ensemble

* We shall find hereafter that the equation

$$\overline{\left(\frac{d\phi}{d\epsilon} - \frac{1}{\Theta}\right)(\epsilon - \bar{\epsilon})} = -1$$

is exact for any value of n greater than 2, and that the equation

$$\overline{\left(\frac{d\phi}{d\epsilon} - \frac{1}{\Theta}\right)^2} = -\overline{\frac{d^2\phi}{d\epsilon^2}}$$

is exact for any value of n greater than 4.

falls within any given limits of energy (ϵ' and ϵ'') is represented by

$$\int_{\epsilon'}^{\epsilon''} e^{\eta+\phi}\, d\epsilon.$$

If we expand η and ϕ in ascending powers of $\epsilon - \epsilon_0$, without going beyond the squares, the probability that the energy falls within the given limits takes the form of the 'law of errors'—

$$e^{\phi_0+\eta_0}\int_{\epsilon'}^{\epsilon''} e^{\left[\left(\frac{d^2\eta}{d\epsilon^2}\right)_0+\left(\frac{d^2\phi}{d\epsilon^2}\right)_0\right]\frac{(\epsilon-\epsilon_0)^2}{2}} d\epsilon. \tag{353}$$

This gives

$$\eta_0 + \phi_0 = \tfrac{1}{2}\log\left[\frac{-1}{2\pi}\left(\frac{d^2\eta}{d\epsilon_2{}^2}\right)_0 - \frac{1}{2\pi}\left(\frac{d^2\phi}{d\epsilon^2}\right)_0\right], \tag{354}$$

and

$$\overline{(\epsilon - \epsilon_0)^2} = \left[-\left(\frac{d^2\eta}{d\epsilon^2}\right)_0 - \left(\frac{d^2\phi}{d\epsilon^2}\right)_0\right]^{-1} \tag{355}$$

We shall have a close approximation in general when the quantities equated in (355) are very small, *i. e.*, when

$$-\left(\frac{d^2\eta}{d\epsilon^2}\right)_0 - \left(\frac{d^2\phi}{d\epsilon^2}\right)_0 \tag{356}$$

is very great. Now when n is very great, $-d^2\phi/d\epsilon^2$ is of the same order of magnitude, and the condition that (356) shall be very great does not restrict very much the nature of the function η.

We may obtain other properties pertaining to average values in a canonical ensemble by the method used for the average of $d\phi/d\epsilon$. Let u be any function of the energy, either alone or with Θ and the external coördinates. The average value of u in the ensemble is determined by the equation

$$\bar{u} = \int_{V=0}^{\epsilon=\infty} u\, e^{\frac{\psi-\epsilon}{\Theta}+\phi}\, d\epsilon. \tag{357}$$

Now we have identically

$$\int_{V=0}^{\epsilon=\infty}\left(\frac{du}{d\epsilon}-\frac{u}{\Theta}+u\,\frac{d\phi}{d\epsilon}\right)e^{\frac{\psi-\epsilon}{\Theta}+\phi}\,d\epsilon=\left[u\,e^{\frac{\psi-\epsilon}{\Theta}+\phi}\right]_{V=0}^{\epsilon=\infty} \tag{358}$$

Therefore, by the preceding equation

$$\overline{\frac{du}{d\epsilon}}-\frac{\overline{u}}{\Theta}+\overline{u\,\frac{d\phi}{d\epsilon}}=\left[u\,e^{\frac{\psi-\epsilon}{\Theta}+\phi}\right]_{V=0}^{\epsilon=\infty} \qquad *(359)$$

If we set $u=1$, (a value which need not be excluded,) the second member of this equation vanishes, as shown on page 101, if $n>2$, and we get

$$\overline{\frac{d\phi}{d\epsilon}}=\frac{1}{\Theta}, \tag{360}$$

as before. It is evident from the same considerations that the second member of (359) will always vanish if $n>2$, unless u becomes infinite at one of the limits, in which case a more careful examination of the value of the expression will be necessary. To facilitate the discussion of such cases, it will be convenient to introduce a certain limitation in regard to the nature of the system considered. We have necessarily supposed, in all our treatment of systems canonically distributed, that the system considered was such as to be capable of the canonical distribution with the given value of the modulus. We shall now suppose that the system is such as to be capable of a canonical distribution with any (finite)† modulus. Let us see what cases we exclude by this last limitation.

* A more general equation, which is not limited to ensembles canonically distributed, is

$$\overline{\frac{du}{d\epsilon}}+\overline{u\,\frac{d\eta}{d\epsilon}}+\overline{u\,\frac{d\phi}{d\epsilon}}=\left[u\,e^{\eta+\phi}\right]_{V=0}^{\epsilon=\infty}$$

where η denotes, as usual, the index of probability of phase.

† The term *finite* applied to the modulus is intended to exclude the value zero as well as infinity.

The impossibility of a canonical distribution occurs when the equation

$$e^{-\frac{\psi}{\Theta}} = \int_{V=0}^{\epsilon=\infty} e^{-\frac{\epsilon}{\Theta}+\phi} d\epsilon \tag{361}$$

fails to determine a finite value for ψ. Evidently the equation cannot make ψ an infinite positive quantity, the impossibility therefore occurs when the equation makes $\psi = -\infty$. Now we get easily from (191)

$$d\frac{\psi}{\Theta} = -\frac{\bar{\epsilon}}{\Theta^2} d\Theta.$$

If the canonical distribution is possible for any values of Θ, we can apply this equation so long as the canonical distribution is possible. The equation shows that as Θ is increased (without becoming infinite) $-\psi$ cannot become infinite unless $\bar{\epsilon}$ simultaneously becomes infinite, and that as Θ is decreased (without becoming zero) $-\psi$ cannot become infinite unless simultaneously $\bar{\epsilon}$ becomes an infinite negative quantity. The corresponding cases in thermodynamics would be bodies which could absorb or give out an infinite amount of heat without passing certain limits of temperature, when no external work is done in the positive or negative sense. Such infinite values present no analytical difficulties, and do not contradict the general laws of mechanics or of thermodynamics, but they are quite foreign to our ordinary experience of nature. In excluding such cases (which are certainly not entirely devoid of interest) we do not exclude any which are analogous to any actual cases in thermodynamics.

We assume then that for any finite value of Θ the second member of (361) has a finite value.

When this condition is fulfilled, the second member of (359) will vanish for $u = e^{-\phi} V$. For, if we set $\Theta' = 2\Theta$,

$$e^{-\frac{\epsilon}{\Theta}} V = e^{-\frac{\epsilon}{\Theta}} \int_{V=0}^{\epsilon} e^{\phi} d\epsilon \leqq e^{-\frac{\epsilon}{\Theta'}} \int_{V=0}^{\epsilon} e^{-\frac{\epsilon}{\Theta'}+\phi} d\epsilon \leqq e^{-\frac{\epsilon'}{\Theta'}} e^{-\frac{\psi'}{\Theta'}},$$

where ψ' denotes the value of ψ for the modulus Θ'. Since the last member of this formula vanishes for $\epsilon = \infty$, the less value represented by the first member must also vanish for the same value of ϵ. Therefore the second member of (359), which differs only by a constant factor, vanishes at the upper limit. The case of the lower limit remains to be considered. Now

$$e^{-\frac{\epsilon}{\Theta}} V \leqq \int_{V=0}^{\epsilon} e^{-\frac{\epsilon}{\Theta}+\phi} d\epsilon.$$

The second member of this formula evidently vanishes for the value of ϵ, which gives $V = 0$, whether this be finite or negative infinity. Therefore, the second member of (359) vanishes at the lower limit also, and we have

$$1 - \overline{e^{-\phi} V \frac{d\phi}{d\epsilon}} - \overline{e^{-\phi} \frac{V}{\Theta}} + \overline{e^{-\phi} V \frac{d\phi}{d\epsilon}} = 0,$$

or

$$\overline{e^{-\phi} V} = \Theta. \qquad (362)$$

This equation, which is subject to no restriction in regard to the value of n, suggests a connection or analogy between the function of the energy of a system which is represented by $\epsilon^{-\phi} V$ and the notion of temperature in thermodynamics. We shall return to this subject in Chapter XIV.

If $n > 2$, the second member of (359) may easily be shown to vanish for any of the following values of u viz.: ϕ, e^{ϕ}, ϵ, ϵ^m, where m denotes any positive number. It will also vanish, when $n > 4$, for $u = d\phi/d\epsilon$, and when $n > 2h$ for $u = e^{-\phi}\, d^h V/d\epsilon^h$. When the second member of (359) vanishes, and $n > 2$, we may write

$$\overline{(u - \bar{u})\left(\frac{d\phi}{d\epsilon} - \frac{1}{\Theta}\right)} = \overline{u\frac{d\phi}{d\epsilon}} - \frac{\bar{u}}{\Theta} = -\overline{\frac{du}{d\epsilon}}. \qquad (363)$$

We thus obtain the following equations:

If $n > 2$,

$$\overline{(\phi - \bar{\phi})\left(\frac{d\phi}{d\epsilon} - \frac{1}{\Theta}\right)} = \overline{\phi\frac{d\phi}{d\epsilon}} - \frac{\bar{\phi}}{\Theta} = -\overline{\frac{1}{\Theta}}, \qquad (364)$$

$$\overline{(e^\phi - \overline{e^\phi})\left(\frac{d\phi}{d\epsilon} - \frac{1}{\Theta}\right)} = \overline{e^\phi \frac{d\phi}{d\epsilon}} - \frac{\overline{e^\phi}}{\Theta} = -\overline{e^\phi \frac{d\phi}{d\epsilon}}, \tag{365}$$

or

$$2\,\overline{e^\phi \frac{d\phi}{d\epsilon}} = \overline{e^\phi}\,\overline{\frac{d\phi}{d\epsilon}} = \frac{\overline{e^\phi}}{\Theta}, \tag{366}$$

$$\overline{(\epsilon - \bar{\epsilon})\left(\frac{d\phi}{d\epsilon} - \frac{1}{\Theta}\right)} = \overline{\epsilon \frac{d\phi}{d\epsilon}} - \frac{\bar{\epsilon}}{\Theta} = -1, \tag{*367}$$

$$\overline{(\epsilon^m - \overline{\epsilon^m})\left(\frac{d\phi}{d\epsilon} - \frac{1}{\Theta}\right)} = \overline{\epsilon^m \frac{d\phi}{d\epsilon}} - \frac{\overline{\epsilon^m}}{\Theta} = -m\,\overline{\epsilon^{m-1}}. \tag{368}$$

If $n > 4$,

$$\overline{\left(\frac{d\phi}{d\epsilon} - \frac{1}{\Theta}\right)^2} = \overline{\left(\frac{d\phi}{d\epsilon}\right)^2} - \frac{1}{\Theta^2} = -\overline{\left(\frac{d^2\phi}{d\epsilon^2}\right)}. \tag{†369}$$

If $n > 2h$,

$$\overline{e^{-\phi} \frac{d^h V}{d\epsilon^h} \frac{d\phi}{d\epsilon}} - \frac{1}{\Theta}\,\overline{e^{-\phi} \frac{d^h V}{d\epsilon^h}} = \overline{e^{-\phi} \frac{d^h V}{d\epsilon^h} \frac{d\phi}{d\epsilon}} - \overline{e^{-\phi} \frac{d^{h+1} V}{d\epsilon^{h+1}}}$$

or

$$\overline{e^{-\phi} \frac{d^{h+1} V}{d\epsilon^{h+1}}} = \frac{1}{\Theta}\,\overline{e^{-\phi} \frac{d^h V}{d\epsilon^h}} \tag{370}$$

whence

$$\overline{e^{-\phi} \frac{d^{h+1} V}{d\epsilon^{h+1}}} = \frac{1}{\Theta^h}. \tag{371}$$

Giving h the values 1, 2, 3, etc., we have

$$\overline{\frac{d\phi}{d\epsilon}} = \frac{1}{\Theta}, \quad \text{if } n > 2,$$

$$\overline{\frac{d^2\phi}{d\epsilon^2}} + \overline{\left(\frac{d\phi}{d\epsilon}\right)^2} = \frac{1}{\Theta^2} \quad \text{if } n > 4,$$

as already obtained. Also

$$\overline{\frac{d^3\phi}{d\epsilon^3}} + 3\,\overline{\frac{d^2\phi}{d\epsilon^2}\frac{d\phi}{d\epsilon}} + \overline{\left(\frac{d\phi}{d\epsilon}\right)^3} = \frac{1}{\Theta^3} \quad \text{if } n > 6. \tag{372}$$

* This equation may also be obtained from equations (252) and (321). Compare also equation (349) which was derived by an approximative method.

† Compare equation (350), obtained by an approximative method.

If V_q is a continuous increasing function of ϵ_q, commencing with $V_q = 0$, the average value in a canonical ensemble of any function of ϵ_q, either alone or with the modulus and the external coördinates, is given by equation (275), which is identical with (357) except that ϵ, ϕ, and ψ have the suffix $(\ )_q$. The equation may be transformed so as to give an equation identical with (359) except for the suffixes. If we add the same suffixes to equation (361), the finite value of its members will determine the possibility of the canonical distribution.

From these data, it is easy to derive equations similar to (360), (362)–(372), except that the conditions of their validity must be differently stated. The equation

$$\overline{e^{-\phi_q} V_q} = \Theta$$

requires only the condition already mentioned with respect to V_q. This equation corresponds to (362), which is subject to no restriction with respect to the value of n. We may observe, however, that V will always satisfy a condition similar to that mentioned with respect to V_q.

If V_q satisfies the condition mentioned, and e^{ϕ_q} a similar condition, *i. e.*, if e^{ϕ_q} is a continuous increasing function of ϵ_q, commencing with the value $e^{\phi_q} = 0$, equations will hold similar to those given for the case when $n > 2$, viz., similar to (360), (364)–(368). Especially important is

$$\overline{\frac{d\phi_q}{d\epsilon_q}} = \frac{1}{\Theta}.$$

If V_q, e^{ϕ_q} (or $dV_q/d\epsilon_q$), $d^2V_q/d\epsilon_q^2$ all satisfy similar conditions, we shall have an equation similar to (369), which was subject to the condition $n > 4$. And if $d^3V_q/d\epsilon_q^3$ also satisfies a similar condition, we shall have an equation similar to (372), for which the condition was $n > 6$. Finally, if V_q and h successive differential coefficients satisfy conditions of the kind mentioned, we shall have equations like (370) and (371) for which the condition was $n > 2h$.

These conditions take the place of those given above relating to n. In fact, we might give conditions relating to the differential coefficients of V, similar to those given relating to the differential coefficients of V_q, instead of the conditions relating to n, for the validity of equations (360), (363)–(372). This would somewhat extend the application of the equations.

CHAPTER X.

ON A DISTRIBUTION IN PHASE CALLED MICROCANONICAL IN WHICH ALL THE SYSTEMS HAVE THE SAME ENERGY.

AN important case of statistical equilibrium is that in which all systems of the ensemble have the same energy. We may arrive at the notion of a distribution which will satisfy the necessary conditions by the following process. We may suppose that an ensemble is distributed with a uniform density-in-phase between two limiting values of the energy, ϵ' and ϵ'', and with density zero outside of those limits. Such an ensemble is evidently in statistical equilibrium according to the criterion in Chapter IV, since the density-in-phase may be regarded as a function of the energy. By diminishing the difference of ϵ' and ϵ'', we may diminish the differences of energy in the ensemble. The limit of this process gives us a permanent distribution in which the energy is constant.

We should arrive at the same result, if we should make the density any function of the energy between the limits ϵ' and ϵ'', and zero outside of those limits. Thus, the limiting distribution obtained from the part of a canonical ensemble between two limits of energy, when the difference of the limiting energies is indefinitely diminished, is independent of the modulus, being determined entirely by the energy, and is identical with the limiting distribution obtained from a uniform density between limits of energy approaching the same value.

We shall call the limiting distribution at which we arrive by this process *microcanonical.*

We shall find however, in certain cases, that for certain values of the energy, viz., for those for which e^{ϕ} is infinite, this process fails to define a limiting distribution in any such distinct sense as for other values of the energy. The difficulty is not in the process, but in the nature of the case, being entirely analogous to that which we meet when we try to find a canonical distribution in cases when ψ becomes infinite. We have not regarded such cases as affording true examples of the canonical distribution, and we shall not regard the cases in which e^{ϕ} is infinite as affording true examples of the microcanonical distribution. We shall in fact find as we go on that in such cases our most important formulae become illusory.

The use of formulae relating to a canonical ensemble which contain $e^{\phi}\,d\epsilon$ instead of $dp_1 \ldots dq_n$, as in the preceding chapters, amounts to the consideration of the ensemble as divided into an infinity of microcanonical elements.

From a certain point of view, the microcanonical distribution may seem more simple than the canonical, and it has perhaps been more studied, and been regarded as more closely related to the fundamental notions of thermodynamics. To this last point we shall return in a subsequent chapter. It is sufficient here to remark that analytically the canonical distribution is much more manageable than the microcanonical.

We may sometimes avoid difficulties which the microcanonical distribution presents by regarding it as the result of the following process, which involves conceptions less simple but more amenable to analytical treatment. We may suppose an ensemble distributed with a density proportional to

$$e^{-\frac{(\epsilon-\epsilon')^2}{\omega^2}},$$

where ω and ϵ' are constants, and then diminish indefinitely the value of the constant ω. Here the density is nowhere zero until we come to the limit, but at the limit it is zero for all energies except ϵ'. We thus avoid the analytical complication of discontinuities in the value of the density, which require the use of integrals with inconvenient limits.

In a microcanonical ensemble of systems the energy (ϵ) is constant, but the kinetic energy (ϵ_p) and the potential energy

(ϵ_q) vary in the different systems, subject of course to the condition

$$\epsilon_p + \epsilon_q = \epsilon = \text{constant}. \tag{373}$$

Our first inquiries will relate to the division of energy into these two parts, and to the average values of functions of ϵ_p and ϵ_q.

We shall use the notation $\overline{u}|_\epsilon$ to denote an average value in a microcanonical ensemble of energy ϵ. An average value in a canonical ensemble of modulus Θ, which has hitherto been denoted by $\overline{u}$, we shall in this chapter denote by $\overline{u}|_\Theta$, to distinguish more clearly the two kinds of averages.

The extension-in-phase within any limits which can be given in terms of ϵ_p and ϵ_q may be expressed in the notations of the preceding chapter by the double integral

$$\int\int dV_p \, dV_q$$

taken within those limits. If an ensemble of systems is distributed within those limits with a uniform density-in-phase, the average value in the ensemble of any function (u) of the kinetic and potential energies will be expressed by the quotient of integrals

$$\frac{\int\int u \, dV_p \, dV_q}{\int\int dV_p \, dV_q}.$$

Since $dV_p = e^{\phi_p} d\epsilon_p$, and $d\epsilon_p = d\epsilon$ when ϵ_q is constant, the expression may be written

$$\frac{\int\int u \, e^{\phi_p} \, d\epsilon \, dV_q}{\int\int e^{\phi_p} \, d\epsilon \, dV_q}.$$

To get the average value of u in an ensemble distributed microcanonically with the energy ϵ, we must make the integrations cover the extension-in-phase between the energies ϵ and $\epsilon + d\epsilon$. This gives

$$\overline{u}\Big]_\epsilon = \frac{d\epsilon \displaystyle\int_{V_q=0}^{\epsilon_q=\epsilon} u\, e^{\phi_p}\, dV_q}{d\epsilon \displaystyle\int_{V_q=0}^{\epsilon_q=\epsilon} e^{\phi_p}\, dV_q}$$

But by (299) the value of the integral in the denominator is e^{ϕ}. We have therefore

$$\overline{u}\Big]_\epsilon = e^{-\phi} \int_{V_q=0}^{\epsilon_q=\epsilon} u\, e^{\phi_p}\, dV_q, \tag{374}$$

where e^{ϕ_p} and V_q are connected by equation (373), and u, if given as function of ϵ_p, or of ϵ_p and ϵ_q, becomes in virtue of the same equation a function of ϵ_q alone.

We shall assume that e^{ϕ} has a finite value. If $n > 1$, it is evident from equation (305) that e^{ϕ} is an increasing function of ϵ, and therefore cannot be infinite for one value of ϵ without being infinite for all greater values of ϵ, which would make $-\psi$ infinite.* When $n > 1$, therefore, if we assume that e^{ϕ} is finite, we only exclude such cases as we found necessary to exclude in the study of the canonical distribution. But when $n = 1$, cases may occur in which the canonical distribution is perfectly applicable, but in which the formulae for the microcanonical distribution become illusory, for particular values of ϵ, on account of the infinite value of e^{ϕ}. Such failing cases of the microcanonical distribution *for particular values of the energy* will not prevent us from regarding the canonical ensemble as consisting of an infinity of microcanonical ensembles.†

* See equation (322).

† An example of the failing case of the microcanonical distribution is afforded by a material point, under the influence of gravity, and constrained to remain in a vertical circle. The failing case occurs when the energy is just sufficient to carry the material point to the highest point of the circle.

It will be observed that the difficulty is inherent in the nature of the case, and is quite independent of the mathematical formulae. The nature of the difficulty is at once apparent if we try to distribute a finite number of

From the last equation, with (298), we get

$$\overline{e^{-\phi_p} V_p}\big|_\epsilon = e^{-\phi} \int_{V_q=0}^{\epsilon_q=\epsilon} V_p \, dV_q = e^{-\phi} V. \tag{375}$$

But by equations (288) and (289)

$$e^{-\phi_p} V_p = \frac{2}{n} \epsilon_p. \tag{376}$$

Therefore

$$e^{-\phi} V = \overline{e^{-\phi_p} V_p}\big|_\epsilon = \frac{2}{n} \overline{\epsilon_p}\big|_\epsilon. \tag{377}$$

Again, with the aid of equation (301), we get

$$\overline{\frac{d\phi_p}{d\epsilon_p}}\bigg|_\epsilon = e^{-\phi} \int_{V_q=0}^{\epsilon_q=\epsilon} \frac{d\phi_p}{d\epsilon_p} e^{\phi_p} \, dV_q = \frac{d\phi}{d\epsilon}, \tag{378}$$

if $n > 2$. Therefore, by (289),

$$\frac{d\phi}{d\epsilon} = \overline{\frac{d\phi_p}{d\epsilon_p}}\bigg|_\epsilon = \left(\frac{n}{2} - 1\right) \overline{\epsilon_p^{-1}}\big|_\epsilon, \quad \text{if} \quad n > 2. \tag{379}$$

These results are interesting on account of the relations of the functions $e^{-\phi} V$ and $\frac{d\phi}{d\epsilon}$ to the notion of temperature in thermodynamics, — a subject to which we shall return hereafter. They are particular cases of a general relation easily deduced from equations (306), (374), (288) and (289). We have

$$\frac{d^h V}{d\epsilon^h} = \int_{V_q=0}^{\epsilon_q=\epsilon} \frac{d^h V_p}{d\epsilon_p^h} \, dV_q, \quad \text{if} \quad h < \tfrac{1}{2} n + 1.$$

The equation may be written

$$e^{-\phi} \frac{d^h V}{d\epsilon^h} = e^{-\phi} \int_{V_q=0}^{\epsilon_q=\epsilon} e^{-\phi_p} \frac{d^h V_p}{d\epsilon_p^h} e^{\phi_p} \, dV_q.$$

material points with this particular value of the energy as nearly as possible in statistical equilibrium, or if we ask: What is the probability that a point taken at random from an ensemble in statistical equilibrium with this value of the energy will be found in any specified part of the circle?

We have therefore

$$e^{-\phi}\frac{d^h V}{d\epsilon^h} = \overline{e^{-\phi_p}\frac{d^h V_p}{d\epsilon_p{}^h}}\Big|_\epsilon = \frac{\Gamma(\frac{1}{2}n)}{\Gamma(\frac{1}{2}n - h + 1)}\overline{\epsilon_p{}^{1-h}}\Big|_\epsilon, \tag{380}$$

if $h < \frac{1}{2}n + 1$. For example, when n is even, we may make $h = \frac{1}{2}n$, which gives, with (307),

$$(2\pi)^{\frac{n}{2}} e^{-\phi}(V_q)_{\epsilon_q=\epsilon} = \Gamma(\tfrac{1}{2}n)\,\overline{\epsilon_p{}^{1-\frac{n}{2}}}\Big|_\epsilon. \tag{381}$$

Since any canonical ensemble of systems may be regarded as composed of microcanonical ensembles, if any quantities u and v have the same average values in every microcanonical ensemble, they will have the same values in every canonical ensemble. To bring equation (380) formally under this rule, we may observe that the first member being a function of ϵ is a constant value in a microcanonical ensemble, and therefore identical with its average value. We get thus the general equation

$$\overline{e^{-\phi}\frac{d^h V}{d\epsilon^h}}\Big|_\Theta = \overline{e^{-\phi_p}\frac{d^h V_p}{d\epsilon_p{}^h}}\Big|_\Theta = \frac{\Gamma(\frac{1}{2}n)}{\Gamma(\frac{1}{2}n - h + 1)}\overline{\epsilon_p{}^{1-h}}\Big|_\Theta = \Theta^{1-h}, \tag{382}$$

if $h < \frac{1}{2}n + 1$.* The equations

$$\Theta = \overline{e^{-\phi}V}\Big|_\Theta = \overline{e^{-\phi_p}V_p}\Big|_\Theta = \frac{2}{n}\overline{\epsilon_p}\Big|_\Theta, \tag{383}$$

$$\frac{1}{\Theta} = \overline{\frac{d\phi}{d\epsilon}}\Big|_\Theta = \overline{\frac{d\phi_p}{d\epsilon_p}}\Big|_\Theta = \left(\frac{n}{2} - 1\right)\overline{\epsilon_p{}^{-1}}\Big|_\Theta, \tag{384}$$

may be regarded as particular cases of the general equation. The last equation is subject to the condition that $n > 2$.

The last two equations give for a canonical ensemble, if $n > 2$,

$$\left(1 - \frac{2}{n}\right)\overline{\epsilon_p}\Big|_\Theta\,\overline{\epsilon_p{}^{-1}}\Big|_\Theta = 1. \tag{385}$$

The corresponding equations for a microcanonical ensemble give, if $n > 2$,

$$\left(1 - \frac{2}{n}\right)\overline{\epsilon_p}\Big|_\epsilon\,\overline{\epsilon_p{}^{-1}}\Big|_\epsilon = \frac{d\phi}{d\log V}, \tag{386}$$

* See equation (292).

which shows that $d\phi/d\log V$ approaches the value unity when n is very great.

If a system consists of two parts, having separate energies, we may obtain equations similar in form to the preceding, which relate to the system as thus divided.* We shall distinguish quantities relating to the parts by letters with suffixes, the same letters without suffixes relating to the whole system. The extension-in-phase of the whole system within any given limits of the energies may be represented by the double integral

$$\int\int dV_1\,dV_2$$

taken within those limits, as appears at once from the definitions of Chapter VIII. In an ensemble distributed with uniform density within those limits, and zero density outside, the average value of any function of ϵ_1 and ϵ_2 is given by the quotient

$$\frac{\int\int u\,dV_1\,dV_2}{\int\int dV_1\,dV_2}$$

which may also be written †

$$\frac{\int\int u\,e^{\phi_1}\,d\epsilon\,dV_2}{\int\int e^{\phi_1}\,d\epsilon\,dV_2}$$

If we make the limits of integration ϵ and $\epsilon + d\epsilon$, we get the

* If this condition is rigorously fulfilled, the parts will have no influence on each other, and the ensemble formed by distributing the whole microcanonically is too arbitrary a conception to have a real interest. The principal interest of the equations which we shall obtain will be in cases in which the condition is approximately fulfilled. But for the purposes of a theoretical discussion, it is of course convenient to make such a condition absolute. Compare Chapter IV, pp. 35 ff., where a similar condition is considered in connection with canonical ensembles.

† Where the analytical transformations are identical in form with those on the preceding pages, it does not appear necessary to give all the steps with the same detail.

average value of u in an ensemble in which the whole system is microcanonically distributed in phase, viz.,

$$\overline{u}|_\epsilon = e^{-\phi} \int_{V_2=0}^{\epsilon_2=\epsilon} u\, e^{\phi_1}\, dV_2, \tag{387}$$

where ϕ_1 and V_2 are connected by the equation

$$\epsilon_1 + \epsilon_2 = \text{constant} = \epsilon, \tag{388}$$

and u, if given as function of ϵ_1, or of ϵ_1 and ϵ_2, becomes in virtue of the same equation a function of ϵ_2 alone.*
Thus

$$\overline{e^{-\phi_1} V_1}|_\epsilon = e^{-\phi} \int_{V_2=0}^{\epsilon_2=\epsilon} V_1\, dV_2, \tag{389}$$

$$e^{-\phi}\, V = \overline{e^{-\phi_1} V_1}|_\epsilon = \overline{e^{-\phi_2} V_2}|_\epsilon. \tag{390}$$

This requires a similar relation for canonical averages

$$\Theta = \overline{e^{-\phi}\, V}|_\Theta = \overline{e^{-\phi_1} V_1}|_\Theta = \overline{e^{-\phi_2} V_2}|_\Theta. \tag{391}$$

Again

$$\overline{\frac{d\phi_1}{d\epsilon_1}}\bigg|_\epsilon = e^{-\phi} \int_{V_2=0}^{\epsilon_2=\epsilon} \frac{d\phi_1}{d\epsilon_1}\, e^{\phi_1}\, dV_2. \tag{392}$$

But if $n_1 > 2$, e^{ϕ_1} vanishes for $V_1 = 0$,† and

$$\frac{d}{d\epsilon}\, e^{\phi} = \frac{d}{d\epsilon} \int_{V_2=0}^{\epsilon_2=\epsilon} e^{\phi_1}\, dV_2 = \int_{V_2=0}^{\epsilon_2=\epsilon} \frac{d\phi_1}{d\epsilon_1}\, e^{\phi_1}\, dV_2. \tag{393}$$

Hence, if $n_1 > 2$, and $n_2 > 2$,

$$\frac{d\phi}{d\epsilon} = \overline{\frac{d\phi_1}{d\epsilon_1}}\bigg|_\epsilon = \overline{\frac{d\phi_2}{d\epsilon_2}}\bigg|_\epsilon, \tag{394}$$

* In the applications of the equation (387), we cannot obtain all the results corresponding to those which we have obtained from equation (374), because ϕ_p is a known function of ϵ_p, while ϕ_1 must be treated as an arbitrary function of ϵ_1, or nearly so.

† See Chapter VIII, equations (305) and (316).

and
$$\frac{1}{\Theta} = \left.\overline{\frac{d\phi}{d\epsilon}}\right|_{\Theta} = \left.\overline{\frac{d\phi_1}{d\epsilon_1}}\right|_{\Theta} = \left.\overline{\frac{d\phi_2}{d\epsilon_2}}\right|_{\Theta} \tag{395}$$

We have compared certain functions of the energy of the whole system with average values of similar functions of the kinetic energy of the whole system, and with average values of similar functions of the whole energy of a part of the system. We may also compare the same functions with average values of the kinetic energy of a part of the system.

We shall express the total, kinetic, and potential energies of the whole system by ϵ, ϵ_p, and ϵ_q, and the kinetic energies of the parts by ϵ_{1p} and ϵ_{2p}. These kinetic energies are necessarily separate: we need not make any supposition concerning potential energies. The extension-in-phase within any limits which can be expressed in terms of ϵ_q, ϵ_{1p}, ϵ_{2p} may be represented in the notations of Chapter VIII by the triple integral

$$\iiint dV_{1p}\,dV_{2p}\,dV_q$$

taken within those limits. And if an ensemble of systems is distributed with a uniform density within those limits, the average value of any function of ϵ_q, ϵ_{1p}, ϵ_{2p} will be expressed by the quotient

$$\frac{\iiint u\,dV_{1p}\,dV_{2p}\,dV_q}{\iiint dV_{1p}\,dV_{2p}\,dV_q}$$

or

$$\frac{\iiint u\,e^{\phi_{1p}}\,d\epsilon\,dV_{2p}\,dV_q}{\iiint e^{\phi_{1p}}\,d\epsilon\,dV_{2p}\,dV_q}$$

To get the average value of u for a microcanonical distribution, we must make the limits ϵ and $\epsilon + d\epsilon$. The denominator in this case becomes $e^{\phi}\,d\epsilon$, and we have

$$\overline{u}|_{\epsilon} = e^{-\phi} \int_{V_q=0}^{\epsilon_q=\epsilon} \int_{\epsilon_{2p}=0}^{\epsilon_{2p}=\epsilon-\epsilon_q} u\,e^{\phi_{1p}}\,dV_{2p}\,dV_q, \tag{396}$$

where ϕ_{1p}, V_{2p}, and V_q are connected by the equation

$$\epsilon_{1p} + \epsilon_{2p} + \epsilon_q = \text{constant} = \epsilon.$$

Accordingly

$$\overline{e^{-\phi_{1p}} V_{1p}}\Big|_\epsilon = e^{-\phi} \int\limits_{V_q=0}^{\epsilon_q=\epsilon} \int\limits_{\epsilon_{2p}=0}^{\epsilon_{2p}=\epsilon-\epsilon_q} V_{1p}\, d V_{2p}\, d V_q = e^{-\phi} V, \qquad (397)$$

and we may write

$$e^{-\phi} V = \overline{e^{-\phi_{1p}} V_{1p}}\Big|_\epsilon = \overline{e^{-\phi_{2p}} V_{2p}}\Big|_\epsilon = \frac{2}{n_1} \overline{\epsilon_{1p}}\Big|_\epsilon = \frac{2}{n_2} \overline{\epsilon_{2p}}\Big|_\epsilon, \qquad (398)$$

and

$$\Theta = \overline{e^{-\phi} V}\Big|_\Theta = \overline{e^{-\phi_{1p}} V_{1p}}\Big|_\Theta = \overline{e^{-\phi_{2p}} V_{2p}}\Big|_\Theta = \frac{2}{n_1} \overline{\epsilon_{1p}}\Big|_\Theta = \frac{2}{n_2} \overline{\epsilon_{2p}}\Big|_\Theta. \qquad (399)$$

Again, if $n_1 > 2$,

$$\overline{\frac{d\phi_{1p}}{d\epsilon_{1p}}}\Bigg| = e^{-\phi} \int\limits_{V_q=0}^{\epsilon_q=\epsilon} \int\limits_{\epsilon_{2p}=0}^{\epsilon_{2p}=\epsilon-\epsilon_q} \frac{d\phi_{1p}}{d\epsilon_{1p}} e^{\phi_{1p}}\, d V_{2p}\, d V_q$$

$$= e^{-\phi} \int\limits_{V_q=0}^{\epsilon_q=\epsilon} \frac{d e^{\phi_p}}{d\epsilon_p}\, d V_q = e^{-\phi} \frac{d e^{\phi}}{d\epsilon} = \frac{d\phi}{d\epsilon}. \qquad (400)$$

Hence, if $n_1 > 2$, and $n_2 > 2$,

$$\frac{d\phi}{d\epsilon} = \overline{\frac{d\phi_{1p}}{d\epsilon_{1p}}}\Bigg|_\epsilon = \overline{\frac{d\phi_{2p}}{d\epsilon_{2p}}}\Bigg|_\epsilon = (\tfrac{1}{2} n_1 - 1) \overline{\epsilon_{1p}^{-1}}\Big|_\epsilon = (\tfrac{1}{2} n_2 - 1) \overline{\epsilon_{2p}^{-1}}\Big|_\epsilon, \qquad (401)$$

$$\frac{1}{\Theta} = \overline{\frac{d\phi}{d\epsilon}}\Bigg|_\Theta = \overline{\frac{d\phi_{1p}}{d\epsilon_{1p}}}\Bigg|_\Theta = \overline{\frac{d\phi_{2p}}{d\epsilon_{2p}}}\Bigg|_\Theta = (\tfrac{1}{2} n_1 - 1) \overline{\epsilon_{1p}^{-1}}\Big|_\Theta = (\tfrac{1}{2} n_2 - 1) \overline{\epsilon_{2p}^{-1}}\Big|_\Theta. \qquad (402)$$

We cannot apply the methods employed in the preceding pages to the microcanonical averages of the (generalized) forces A_1, A_2, etc., exerted by a system on external bodies, since these quantities are not functions of the energies, either kinetic or potential, of the whole or any part of the system. We may however use the method described on page 116.

Let us imagine an ensemble of systems distributed in phase according to the index of probability

$$c - \frac{(\epsilon - \epsilon')^2}{\omega^2},$$

where ϵ' is any constant which is a possible value of the energy, except only the least value which is consistent with the values of the external coördinates, and c and ω are other constants. We have therefore

$$\int_{\text{phases}}^{\text{all}} \cdots \int e^{c - \frac{(\epsilon-\epsilon')^2}{\omega^2}} dp_1 \ldots dq_n = 1, \tag{403}$$

or

$$e^{-c} = \int_{\text{phases}}^{\text{all}} \cdots \int e^{-\frac{(\epsilon-\epsilon')^2}{\omega^2}} dp_1 \ldots dq_n, \tag{404}$$

or again

$$e^{-c} = \int_{V=0}^{\epsilon=\infty} e^{-\frac{(\epsilon-\epsilon')^2}{\omega^2} + \phi} d\epsilon. \tag{405}$$

From (404) we have

$$\frac{de^{-c}}{da_1} = \int_{\text{phases}}^{\text{all}} \cdots \int 2 \frac{\epsilon - \epsilon'}{\omega^2} A_1 \, e^{-\frac{(\epsilon-\epsilon')^2}{\omega^2}} dp_1 \ldots dq_n$$

$$= \int_{V=0}^{\epsilon=\infty} 2 \, \frac{\epsilon - \epsilon'}{\omega^2} \overline{A_1}|_\epsilon \, e^{-\frac{(\epsilon-\epsilon')^2}{\omega^2} + \phi} d\epsilon, \tag{406}$$

where $\overline{A_1}|_\epsilon$ denotes the average value of A_1 in those systems of the ensemble which have any same energy ϵ. (This is the same thing as the average value of A_1 in a microcanonical ensemble of energy ϵ.) The validity of the transformation is evident, if we consider separately the part of each integral which lies between two infinitesimally differing limits of energy. Integrating by parts, we get

$$\frac{de^{-c}}{da_1} = -\left[\overline{A_1}|_\epsilon \, e^{-\frac{(\epsilon-\epsilon')^2}{\omega^2}+\phi}\right]_{V=0}^{\epsilon=\infty}$$

$$+\int_{V=0}^{\epsilon=\infty}\left(\frac{d\overline{A_1}|_\epsilon}{d\epsilon}+\overline{A_1}|_\epsilon\frac{d\phi}{d\epsilon}\right)e^{-\frac{(\epsilon-\epsilon')}{\omega^2}+\phi}\,d\epsilon. \quad (407)$$

Differentiating (405), we get

$$\frac{de^{-c}}{da_1} = \int_{V=0}^{\epsilon=\infty}\frac{d\phi}{da_1}e^{-\frac{(\epsilon-\epsilon')^2}{\omega^2}+\phi}\,d\epsilon - \left(e^{-\frac{(\epsilon-\epsilon')^2}{\omega^2}+\phi}\frac{d\epsilon_a}{da_1}\right)_{V=0} \quad (408)$$

where ϵ_a denotes the least value of ϵ consistent with the external coördinates. The last term in this equation represents the part of de^{-c}/da_1 which is due to the variation of the lower limit of the integral. It is evident that the expression in the brackets will vanish at the upper limit. At the lower limit, at which $\epsilon_p = 0$, and ϵ_q has the least value consistent with the external coördinates, the average sign on $\overline{A_1}|_\epsilon$ is superfluous, as there is but one value of A_1 which is represented by $-d\epsilon_a/da_1$. Exceptions may indeed occur for particular values of the external coördinates, at which $d\epsilon_a/da_1$ receive a finite increment, and the formula becomes illusory. Such particular values we may for the moment leave out of account. The last term of (408) is therefore equal to the first term of the second member of (407). (We may observe that both vanish when $n > 2$ on account of the factor e^ϕ.)

We have therefore from these equations

$$\int_{V=0}^{\epsilon=\infty}\left(\frac{d\overline{A_1}|_\epsilon}{d\epsilon}+\overline{A_1}|_\epsilon\frac{d\phi}{d\epsilon}\right)e^{-\frac{(\epsilon-\epsilon')^2}{\omega^2}+\phi}\,d\epsilon = \int_{V=0}^{\epsilon=\infty}\frac{d\phi}{da_1}e^{-\frac{(\epsilon-\epsilon')^2}{\omega^2}+\phi}\,d\epsilon,$$

or

$$\int_{V=0}^{\epsilon=\infty}\left(\frac{d\overline{A_1}|_\epsilon}{d\epsilon}+\overline{A_1}|_\epsilon\frac{d\phi}{d\epsilon}-\frac{d\phi}{da_1}\right)e^{c-\frac{(\epsilon-\epsilon')^2}{\omega^2}+\phi}\,d\epsilon = 0. \quad (409)$$

That is: the average value in the ensemble of the quantity represented by the principal parenthesis is zero. This must be true for any value of ω. If we diminish ω, the average value of the parenthesis at the limit when ω vanishes becomes identical with the value for $\epsilon = \epsilon'$. But this may be any value of the energy, except the least possible. We have therefore

$$\frac{d\overline{A_1}|_\epsilon}{d\epsilon} + \overline{A_1}|_\epsilon \frac{d\phi}{d\epsilon} - \frac{d\phi}{da_1} = 0, \tag{410}$$

unless it be for the least value of the energy consistent with the external coördinates, or for particular values of the external coördinates. But the value of any term of this equation as determined for particular values of the energy and of the external coördinates is not distinguishable from its value as determined for values of the energy and external coördinates indefinitely near those particular values. The equation therefore holds without limitation. Multiplying by e^ϕ, we get

$$e^\phi \frac{d\overline{A_1}|_\epsilon}{d\epsilon} + \overline{A_1}|_\epsilon\, e^\phi \frac{d\phi}{d\epsilon} = e^\phi \frac{d\phi}{da_1} = \frac{de^\phi}{da_1} = \frac{d^2 V}{da_1\, d\epsilon}. \tag{411}$$

The integral of this equation is

$$\overline{A_1}|_\epsilon e^\phi = \frac{dV}{da_1} + F_1, \tag{412}$$

where F_1 is a function of the external coördinates. We have an equation of this form for each of the external coördinates. This gives, with (266), for the complete value of the differential of V

$$dV = e^\phi d\epsilon + (e^\phi \overline{A_1}|_\epsilon - F_1)\, da_1 + (e^\phi \overline{A_2}|_\epsilon - F_2)\, da_2 + \text{etc.}, \tag{413}$$

or

$$dV = e^\phi (d\epsilon + \overline{A_1}|_\epsilon da_1 + \overline{A_2}|_\epsilon da_2 + \text{etc.}) - F_1 da_1 - F_2 da_2 - \text{etc.} \tag{414}$$

To determine the values of the functions F_1, F_2, etc., let us suppose a_1, a_2, etc. to vary arbitrarily, while ϵ varies so as always to have the least value consistent with the values of the external coördinates. This will make $V = 0$, and $dV = 0$. If $n < 2$, we shall have also $e^\phi = 0$, which will give

$$F_1 = 0, \quad F_2 = 0, \quad \text{etc.} \tag{415}$$

The result is the same for any value of n. For in the variations considered the kinetic energy will be constantly zero, and the potential energy will have the least value consistent with the external coördinates. The condition of the least possible potential energy may limit the ensemble at each instant to a single configuration, or it may not do so; but in any case the values of A_1, A_2, etc. will be the same at each instant for all the systems of the ensemble,* and the equation

$$d\epsilon + A_1 da_1 + A_2 da_2 + \text{etc.} = 0$$

will hold for the variations considered. Hence the functions F_1, F_2, etc. vanish in any case, and we have the equation

$$dV = e^{\phi}\, d\epsilon + e^{\phi}\, \overline{A_1}\big]_\epsilon da_1 + e^{\phi}\, \overline{A_2}\big]_\epsilon da_2 + \text{etc.}, \qquad (416)$$

or

$$d \log V = \frac{d\epsilon + \overline{A_1}\big]_\epsilon da_1 + \overline{A_2}\big]_\epsilon da_2 + \text{etc.}}{e^{-\phi}\, V}, \qquad (417)$$

or again

$$d\epsilon = e^{-\phi}\, V\, d \log V - \overline{A_1}\big]_\epsilon da_1 - \overline{A_2}\big]_\epsilon da_2 - \text{etc.} \qquad (418)$$

It will be observed that the two last equations have the form of the fundamental differential equations of thermodynamics, $e^{-\phi} V$ corresponding to temperature and $\log V$ to entropy. We have already observed properties of $e^{-\phi} V$ suggestive of an analogy with temperature.† The significance of these facts will be discussed in another chapter.

The two last equations might be written more simply

$$dV = \frac{d\epsilon + \overline{A_1}\big]_\epsilon da_1 + \overline{A_2}\big]_\epsilon da_2 + \text{etc.}}{e^{-\phi}},$$

$$d\epsilon = e^{-\phi}\, dV - \overline{A_1}\big]_\epsilon da_1 - \overline{A_2}\big]_\epsilon da_2 - \text{etc.},$$

and still have the form analogous to the thermodynamic equations, but $e^{-\phi}$ has nothing like the analogies with temperature which we have observed in $e^{-\phi} V$.

* This statement, as mentioned before, may have exceptions for particular values of the external coördinates. This will not invalidate the reasoning, which has to do with varying values of the external coördinates.

† See Chapter IX, page 111; also this chapter, page 119.

CHAPTER XI.

MAXIMUM AND MINIMUM PROPERTIES OF VARIOUS DISTRIBUTIONS IN PHASE.

In the following theorems we suppose, as always, that the systems forming an ensemble are identical in nature and in the values of the external coördinates, which are here regarded as constants.

Theorem I. If an ensemble of systems is so distributed in phase that the index of probability is a function of the energy, the average value of the index is less than for any other distribution in which the distribution in energy is unaltered.

Let us write η for the index which is a function of the energy, and $\eta + \Delta\eta$ for any other which gives the same distribution in energy. It is to be proved that

$$\int_{\text{phases}}^{\text{all}} \cdots \int (\eta + \Delta\eta)\, e^{\eta+\Delta\eta}\, dp_1 \ldots dq_n > \int_{\text{phases}}^{\text{all}} \cdots \int \eta\, e^{\eta}\, dp_1 \ldots dq_n, \quad (419)$$

where η is a function of the energy, and $\Delta\eta$ a function of the phase, which are subject to the conditions that

$$\int_{\text{phases}}^{\text{all}} \cdots \int e^{\eta+\Delta\eta}\, dp_1 \ldots dq_n = \int_{\text{phases}}^{\text{all}} \cdots \int e^{\eta}\, dp_1 \ldots dq_n = 1, \quad (420)$$

and that for any value of the energy (ϵ')

$$\int_{\epsilon=\epsilon'}^{\epsilon=\epsilon'+d\epsilon'} \cdots \int e^{\eta+\Delta\eta}\, dp_1 \ldots dq_n = \int_{\epsilon=\epsilon'}^{\epsilon=\epsilon'+d\epsilon'} \cdots \int e^{\eta} dp_1 \ldots dq_n. \quad (421)$$

Equation (420) expresses the general relations which η and $\eta + \Delta\eta$ must satisfy in order to be indices of any distributions, and (421) expresses the condition that they give the same distribution in energy.

Since η is a function of the energy, and may therefore be regarded as a constant within the limits of integration of (421), we may multiply by η under the integral sign in both members, which gives

$$\int_{\epsilon=\epsilon'}^{\epsilon=\epsilon'+d\epsilon'} \cdots \int \eta e^{\eta+\Delta\eta} \, dp_1 \ldots dq_n = \int_{\epsilon=\epsilon'}^{\epsilon=\epsilon'+d\epsilon'} \cdots \int \eta e^{\eta} \, dp_1 \ldots dq_n.$$

Since this is true within the limits indicated, and for every value of ϵ', it will be true if the integrals are taken for all phases. We may therefore cancel the corresponding parts of (419), which gives

$$\int_{\text{phases}}^{\text{all}} \cdots \int \Delta\eta e^{\eta+\Delta\eta} \, dp_1 \ldots dq_n > 0. \qquad (422)$$

But by (420) this is equivalent to

$$\int_{\text{phases}}^{\text{all}} \cdots \int (\Delta\eta e^{\Delta\eta} + 1 - e^{\Delta\eta}) e^{\eta} \, dp_1 \ldots dq_n > 0. \qquad (423)$$

Now $\Delta\eta \, e^{\Delta\eta} + 1 - e^{\Delta\eta}$ is a decreasing function of $\Delta\eta$ for negative values of $\Delta\eta$, and an increasing function of $\Delta\eta$ for positive values of $\Delta\eta$. It vanishes for $\Delta\eta = 0$. The expression is therefore incapable of a negative value, and can have the value 0 only for $\Delta\eta = 0$. The inequality (423) will hold therefore unless $\Delta\eta = 0$ for all phases. The theorem is therefore proved.

Theorem II. If an ensemble of systems is canonically distributed in phase, the average index of probability is less than in any other distribution of the ensemble having the same average energy.

For the canonical distribution let the index be $(\psi - \epsilon)/\Theta$, and for another having the same average energy let the index be $(\psi - \epsilon)/\Theta + \Delta\eta$, where $\Delta\eta$ is an arbitrary function of the phase subject only to the limitation involved in the notion of the index, that

$$\int_{\text{phases}}^{\text{all}} \cdots \int e^{\frac{\psi-\epsilon}{\Theta}+\Delta\eta} dp_1 \ldots dq_n = \int_{\text{phases}}^{\text{all}} \cdots \int e^{\frac{\psi-\epsilon}{\Theta}} dp_1 \ldots dq_n = 1, \tag{424}$$

and to that relating to the constant average energy, that

$$\int_{\text{phases}}^{\text{all}} \cdots \int \epsilon\, e^{\frac{\psi-\epsilon}{\Theta}+\Delta\eta} dp_1 \ldots dq_n = \int_{\text{phases}}^{\text{all}} \cdots \int \epsilon\, e^{\frac{\psi-\epsilon}{\Theta}} dp_1 \ldots dq_n. \tag{425}$$

It is to be proved that

$$\int_{\text{phases}}^{\text{all}} \cdots \int \left(\frac{\psi}{\Theta} - \frac{\epsilon}{\Theta} + \Delta\eta\right) e^{\frac{\psi-\epsilon}{\Theta}+\Delta\eta} dp_1 \ldots dq_n >$$

$$\int_{\text{phases}}^{\text{all}} \cdots \int \left(\frac{\psi}{\Theta} - \frac{\epsilon}{\Theta}\right) e^{\frac{\psi-\epsilon}{\Theta}} dp_1 \ldots dq_n. \tag{426}$$

Now in virtue of the first condition (424) we may cancel the constant term ψ / Θ in the parentheses in (426), and in virtue of the second condition (425) we may cancel the term ϵ / Θ. The proposition to be proved is thus reduced to

$$\int_{\text{phases}}^{\text{all}} \cdots \int \Delta\eta\, e^{\frac{\psi-\epsilon}{\Theta}+\Delta\eta} dp_1 \ldots dq_n > 0,$$

which may be written, in virtue of the condition (424),

$$\int_{\text{phases}}^{\text{all}} \cdots \int (\Delta\eta\, e^{\Delta\eta} + 1 - e^{\Delta\eta})\, e^{\frac{\psi-\epsilon}{\Theta}} dp_1 \ldots dq_n > 0. \tag{427}$$

In this form its truth is evident for the same reasons which applied to (423).

Theorem III. If Θ is any positive constant, the average value in an ensemble of the expression $\eta + \epsilon / \Theta$ (η denoting as usual the index of probability and ϵ the energy) is less when the ensemble is distributed canonically with modulus Θ, than for any other distribution whatever.

In accordance with our usual notation let us write $(\psi - \epsilon) / \Theta$ for the index of the canonical distribution. In any other distribution let the index be $(\psi - \epsilon) / \Theta + \Delta\eta$.

In the canonical ensemble $\eta + \epsilon / \Theta$ has the constant value ψ / Θ; in the other ensemble it has the value $\psi / \Theta + \Delta\eta$. The proposition to be proved may therefore be written

$$\frac{\psi}{\Theta} < \int \underset{\text{phases}}{\overset{\text{all}}{\cdots}} \int \left(\frac{\psi}{\Theta} + \Delta\eta \right) e^{\frac{\psi - \epsilon}{\Theta} + \Delta\eta} dp_1 \ldots dq_n, \tag{428}$$

where

$$\int \underset{\text{phases}}{\overset{\text{all}}{\cdots}} \int e^{\frac{\psi - \epsilon}{\Theta} + \Delta\eta} dp_1 \ldots dq_n = \int \underset{\text{phases}}{\overset{\text{all}}{\cdots}} \int e^{\frac{\psi - \epsilon}{\Theta}} dp_1 \ldots dq_n = 1. \tag{429}$$

In virtue of this condition, since ψ / Θ is constant, the proposition to be proved reduces to

$$0 < \int \underset{\text{phases}}{\overset{\text{all}}{\cdots}} \int \Delta\eta \, e^{\frac{\psi - \epsilon}{\Theta} + \Delta\eta} dq_1 \ldots dp_n, \tag{430}$$

where the demonstration may be concluded as in the last theorem.

If we should substitute for the energy in the preceding theorems any other function of the phase, the theorems, *mutatis mutandis*, would still hold. On account of the unique importance of the energy as a function of the phase, the theorems as given are especially worthy of notice. When the case is such that other functions of the phase have important properties relating to statistical equilibrium, as described in Chapter IV,* the three following theorems, which are generalizations of the preceding, may be useful. It will be sufficient to give them without demonstration, as the principles involved are in no respect different.

Theorem IV. If an ensemble of systems is so distributed in phase that the index of probability is any function of F_1, F_2, etc., (these letters denoting functions of the phase,) the average value of the index is less than for any other distribution in phase in which the distribution with respect to the functions F_1, F_2, etc. is unchanged.

* See pages 37–41.

Theorem V. If an ensemble of systems is so distributed in phase that the index of probability is a linear function of F_1, F_2, etc., (these letters denoting functions of the phase,) the average value of the index is less than for any other distribution in which the functions F_1, F_2, etc. have the same average values.

Theorem VI. The average value in an ensemble of systems of $\eta + F$ (where η denotes as usual the index of probability and F any function of the phase) is less when the ensemble is so distributed that $\eta + F$ is constant than for any other distribution whatever.

Theorem VII. If a system which in its different phases constitutes an ensemble consists of two parts, and we consider the average index of probability for the whole system, and also the average indices for each of the parts taken separately, the sum of the average indices for the parts will be either less than the average index for the whole system, or equal to it, but cannot be greater. The limiting case of equality occurs when the distribution in phase of each part is independent of that of the other, and only in this case.

Let the coördinates and momenta of the whole system be $q_1 \ldots q_n, p_1, \ldots p_n$, of which $q_1 \ldots q_m\ p_1, \ldots p_m$ relate to one part of the system, and $q_{m+1}, \ldots q_n, p_{m+1}, \ldots p_n$ to the other. If the index of probability for the whole system is denoted by η, the probability that the phase of an unspecified system lies within any given limits is expressed by the integral

$$\int \ldots \int e^\eta dp_1 \ldots dq_n \qquad (431)$$

taken for those limits. If we set

$$\int \ldots \int e^\eta\, dp_{m+1} \ldots dp_n dq_{m+1} \ldots dq_n = e^{\eta_1}, \qquad (432)$$

where the integrations cover all phases of the second system, and

$$\int \ldots \int e^\eta\ dp_1 \ldots dp_m dq_1 \ldots dq_m = e^{\eta_2}, \qquad (433)$$

where the integrations cover all phases of the first system, the integral (431) will reduce to the form

$$\int \ldots \int e^{\eta_1}\, dp_1 \ldots dp_m\, dq_1 \ldots dq_m, \tag{434}$$

when the limits can be expressed in terms of the coördinates and momenta of the first part of the system. The same integral will reduce to

$$\int \ldots \int e^{\eta_2}\, dp_{m+1} \ldots dp_n\, dq_{m+1} \ldots dq_n, \tag{435}$$

when the limits can be expressed in terms of the coördinates and momenta of the second part of the system. It is evident that η_1 and η_2 are the indices of probability for the two parts of the system taken separately.

The main proposition to be proved may be written

$$\int \ldots \int \eta_1\, e^{\eta_1}\, dp_1 \ldots dq_m + \int \ldots \int \eta_2\, e^{\eta_2}\, dp_{m+1} \ldots dq_n \leqq \int \ldots \int \eta\, e^{\eta}\, dp_1 \ldots dq_n, \tag{436}$$

where the first integral is to be taken over all phases of the first part of the system, the second integral over all phases of the second part of the system, and the last integral over all phases of the whole system. Now we have

$$\int \ldots \int e^{\eta}\, dp_1 \ldots dq_n = 1, \tag{437}$$

$$\int \ldots \int e^{\eta_1} dp_1 \ldots dq_m = 1, \tag{438}$$

and

$$\int \ldots \int e^{\eta_2} dp_{m+1} \ldots dq_n = 1, \tag{439}$$

where the limits cover in each case all the phases to which the variables relate. The two last equations, which are in themselves evident, may be derived by partial integration from the first.

It appears from the definitions of η_1 and η_2 that (436) may also be written

$$\int \cdots \int \eta_1 e^{\eta} dp_1 \ldots dq_n + \int \cdots \int \eta_2 e^{\eta} dp_1 \ldots dq_n \leqq \int \cdots \int \eta e^{\eta} dp_1 \ldots dq_n, \tag{440}$$

or

$$\int \cdots \int (\eta - \eta_1 - \eta_2) e^{\eta} dp_1 \ldots dq_n \geqq 0,$$

where the integrations cover all phases. Adding the equation

$$\int \cdots \int e^{\eta_1 + \eta_2} dp_1 \ldots dq_n = 1, \tag{441}$$

which we get by multiplying (438) and (439), and subtracting (437), we have for the proposition to be proved

$$\underset{\text{phases}}{\overset{\text{all}}{\int \cdots \int}} [(\eta - \eta_1 - \eta_2) e^{\eta} + e^{\eta_1 + \eta_2} - e^{\eta}] dp_1 \ldots dq_n \geqq 0. \tag{442}$$

Let

$$u = \eta - \eta_1 - \eta_2. \tag{443}$$

The main proposition to be proved may be written

$$\underset{\text{phases}}{\overset{\text{all}}{\int \cdots \int}} (u e^{u} + 1 - e^{u}) e^{\eta_1 + \eta_2} dp_1 \ldots dq_n \geqq 0. \tag{444}$$

This is evidently true since the quantity in the parenthesis is incapable of a negative value.* Moreover the sign = can hold only when the quantity in the parenthesis vanishes for all phases, *i. e.*, when $u = 0$ for all phases. This makes $\eta = \eta_1 + \eta_2$ for all phases, which is the analytical condition which expresses that the distributions in phase of the two parts of the system are independent.

Theorem VIII. If two or more ensembles of systems which are identical in nature, but may be distributed differently in phase, are united to form a single ensemble, so that the probability-coefficient of the resulting ensemble is a linear function

* See Theorem I, where this is proved of a similar expression.

of the probability-coefficients of the original ensembles, the average index of probability of the resulting ensemble cannot be greater than the same linear function of the average indices of the original ensembles. It can be equal to it only when the original ensembles are similarly distributed in phase.

Let P_1, P_2, etc. be the probability-coefficients of the original ensembles, and P that of the ensemble formed by combining them; and let N_1, N_2, etc. be the numbers of systems in the original ensembles. It is evident that we shall have

$$P = c_1 P_1 + c_2 P_2 + \text{etc.} = \Sigma (c_1 P_1), \tag{445}$$

where

$$c_1 = \frac{N_1}{\Sigma N_1}, \quad c_2 = \frac{N_2}{\Sigma N_1}, \quad \text{etc.} \tag{446}$$

The main proposition to be proved is that

$$\int_{\text{phases}}^{\text{all}} \cdots \int P \log P \, dp_1 \ldots dq_n \leqq \Sigma \left[c_1 \int_{\text{phases}}^{\text{all}} \cdots \int P_1 \log P_1 \, dp_1 \ldots dq_n \right] \tag{447}$$

or

$$\int_{\text{phases}}^{\text{all}} \cdots \int [\Sigma (c_1 P_1 \log P_1) - P \log P] \, dp_1 \ldots dq_n \geqq 0. \tag{448}$$

If we set

$$Q_1 = P_1 \log P_1 - P_1 \log P - P_1 + P$$

Q_1 will be positive, except when it vanishes for $P_1 = P$. To prove this, we may regard P_1 and P as any positive quantities. Then

$$\left(\frac{dQ_1}{dP_1} \right)_P = \log P_1 - \log P,$$

$$\left(\frac{d^2 Q_1}{dP_1^2} \right)_P = \frac{1}{P_1}.$$

Since Q_1 and dQ_1/dP_1 vanish for $P_1 = P$, and the second differential coefficient is always positive, Q_1 must be positive except when $P_1 = P$. Therefore, if Q_2, etc. have similar definitions,

$$\Sigma (c_1 Q_1) \geqq 0. \tag{449}$$

But since
$$\Sigma\,(c_1\,P_1) = P$$
and
$$\Sigma\,c_1 = 1,$$
$$\Sigma\,(c_1\,Q_1) = \Sigma\,(c_1\,P_1 \log P_1) - P \log P. \qquad (450)$$
This proves (448), and shows that the sign = will hold only when
$$P_1 = P, \quad P_2 = P, \quad \text{etc.}$$
for all phases, *i. e.*, only when the distribution in phase of the original ensembles are all identical.

Theorem IX. A uniform distribution of a given number of systems within given limits of phase gives a less average index of probability of phase than any other distribution.

Let η be the constant index of the uniform distribution, and $\eta + \Delta\eta$ the index of some other distribution. Since the number of systems within the given limits is the same in the two distributions we have
$$\int \ldots \int e^{\eta+\Delta\eta}\, dp_1 \ldots dq_n = \int \ldots \int e^{\eta}\, dp_1 \ldots dq_n, \qquad (451)$$
where the integrations, like those which follow, are to be taken within the given limits. The proposition to be proved may be written
$$\int \ldots \int (\eta + \Delta\eta)\, e^{\eta+\Delta\eta}\, dp_1 \ldots dq_n > \int \ldots \int \eta\, e^{\eta}\, dp_1 \ldots dq_n, \quad (452)$$
or, since η is constant,
$$\int \ldots \int (\eta + \Delta\eta)\, e^{\Delta\eta}\, dp_1 \ldots dq_n > \int \ldots \int \eta\, dp_1 \ldots dq_n. \qquad (453)$$
In (451) also we may cancel the constant factor e^{η}, and multiply by the constant factor $(\eta + 1)$. This gives
$$\int \ldots \int (\eta + 1)\, e^{\Delta\eta}\, dp_1 \ldots dq_n = \int \ldots \int (\eta + 1)\, dp_1 \ldots dq_n.$$
The subtraction of this equation will not alter the inequality to be proved, which may therefore be written
$$\int \ldots \int (\Delta\eta - 1)\, e^{\Delta\eta}\, dp_1 \ldots dq_n > \int \ldots \int -\, dp_1 \ldots dq_n$$

or
$$\int \ldots \int (\Delta\eta\, e^{\Delta\eta} - e^{\Delta\eta} + 1)\, dp_1 \ldots dq_n > 0. \qquad (454)$$

Since the parenthesis in this expression represents a positive value, except when it vanishes for $\Delta\eta = 0$, the integral will be positive unless $\Delta\eta$ vanishes everywhere within the limits, which would make the difference of the two distributions vanish. The theorem is therefore proved.

CHAPTER XII.

ON THE MOTION OF SYSTEMS AND ENSEMBLES OF SYSTEMS THROUGH LONG PERIODS OF TIME.

An important question which suggests itself in regard to any case of dynamical motion is whether the system considered will return in the course of time to its initial phase, or, if it will not return exactly to that phase, whether it will do so to any required degree of approximation in the course of a sufficiently long time. To be able to give even a partial answer to such questions, we must know something in regard to the dynamical nature of the system. In the following theorem, the only assumption in this respect is such as we have found necessary for the existence of the canonical distribution.

If we imagine an ensemble of identical systems to be distributed with a uniform density throughout any finite extension-in-phase, the number of the systems which leave the extension-in-phase and will not return to it in the course of time is less than any assignable fraction of the whole number; *provided*, that the total extension-in-phase for the systems considered between two limiting values of the energy is finite, these limiting values being less and greater respectively than any of the energies of the first-mentioned extension-in-phase.

To prove this, we observe that at the moment which we call initial the systems occupy the given extension-in-phase. It is evident that some systems must leave the extension immediately, unless all remain in it forever. Those systems which leave the extension at the first instant, we shall call the *front* of the ensemble. It will be convenient to speak of this front as *generating* the extension-in-phase through which it passes in the course of time, as in geometry a surface is said to generate the volume through which it passes. In equal times the front generates equal extensions in phase. This is an immediate consequence of the principle of *conservation of extension-in-phase*, unless indeed we prefer to consider it as a slight variation in the expression of that principle. For in two equal short intervals of time let the extensions generated be A and B. (We make the intervals short simply to avoid the complications in the enunciation or interpretation of the principle which would arise when the same extension-in-phase is generated more than once in the interval considered.) Now if we imagine that at a given instant systems are distributed throughout the extension A, it is evident that the same systems will after a certain time occupy the extension B, which is therefore equal to A in virtue of the principle cited. The front of the ensemble, therefore, goes on generating equal extensions in equal times. But these extensions are included in a finite extension, viz., that bounded by certain limiting values of the energy. Sooner or later, therefore, the front must generate phases which it has before generated. Such second generation of the same phases must commence with the initial phases. Therefore a portion at least of the front must return to the original extension-in-phase. The same is of course true of the portion of the ensemble which follows that portion of the front through the same phases at a later time.

It remains to consider how large the portion of the ensemble is, which will return to the original extension-in-phase. There can be no portion of the given extension-in-phase, the systems of which leave the extension and do not return. For we can prove for any portion of the extension as for the whole, that at least a portion of the systems leaving it will return.

We may divide the given extension-in-phase into parts as follows. There may be parts such that the systems within them will never pass out of them. These parts may indeed constitute the whole of the given extension. But if the given extension is very small, these parts will in general be non-existent. There may be parts such that systems within them will all pass out of the given extension and all return within it. The whole of the given extension-in-phase is made up of parts of these two kinds. This does not exclude the possibility of phases on the boundaries of such parts, such that systems starting with those phases would leave the extension and never return. But in the supposed distribution of an ensemble of systems with a uniform density-in-phase, such systems would not constitute any assignable fraction of the whole number.

These distinctions may be illustrated by a very simple example. If we consider the motion of a rigid body of which one point is fixed, and which is subject to no forces, we find three cases. (1) The motion is periodic. (2) The system will never return to its original phase, but will return infinitely near to it. (3) The system will never return either exactly or approximately to its original phase. But if we consider any extension-in-phase, however small, a system leaving that extension will return to it except in the case called by Poinsot 'singular,' viz., when the motion is a rotation about an axis lying in one of two planes having a fixed position relative to the rigid body. But all such phases do not constitute any true *extension-in-phase* in the sense in which we have defined and used the term.*

In the same way it may be proved that the systems in a canonical ensemble which at a given instant are contained within any finite extension-in-phase will in general return to

* An ensemble of systems distributed in phase is a less simple and elementary conception than a single system. But by the consideration of suitable ensembles instead of single systems, we may get rid of the inconvenience of having to consider exceptions formed by particular cases of the integral equations of motion, these cases simply disappearing when the ensemble is substituted for the single system as a subject of study. This is especially true when the ensemble is distributed, as in the case called canonical, throughout an extension-in-phase. In a less degree it is true of the microcanonical ensemble, which does not occupy any extension-in-phase, (in the sense in which we have used the term,) although it is convenient to regard it as a limiting case with respect to ensembles which do, as we thus gain for the subject some part of the analytical simplicity which belongs to the theory of ensembles which occupy true extensions-in-phase.

that extension-in-phase, if they leave it, the exceptions, *i. e.*, the number which pass out of the extension-in-phase and do not return to it, being less than any assignable fraction of the whole number. In other words, the probability that a system taken at random from the part of a canonical ensemble which is contained within any given extension-in-phase, will pass out of that extension and not return to it, is zero.

A similar theorem may be enunciated with respect to a microcanonical ensemble. Let us consider the fractional part of such an ensemble which lies within any given limits of phase. This fraction we shall denote by F. It is evidently constant in time since the ensemble is in statistical equilibrium. The systems within the limits will not in general remain the same, but some will pass out in each unit of time while an equal number come in. Some may pass out never to return within the limits. But the number which in any time however long pass out of the limits never to return will not bear any finite ratio to the number within the limits at a given instant. For, if it were otherwise, let f denote the fraction representing such ratio for the time T. Then, in the time T, the number which pass out never to return will bear the ratio fF to the whole number in the ensemble, and in a time exceeding $T/(fF)$ the number which pass out of the limits never to return would exceed the total number of systems in the ensemble. The proposition is therefore proved.

This proof will apply to the cases before considered, and may be regarded as more simple than that which was given. It may also be applied to any true case of statistical equilibrium. By a true case of statistical equilibrium is meant such as may be described by giving the general value of the probability that an unspecified system of the ensemble is contained within any given limits of phase.*

* An ensemble in which the systems are material points constrained to move in vertical circles, with just enough energy to carry them to the highest points, cannot afford a true example of statistical equilibrium. For any other value of the energy than the critical value mentioned, we might

Let us next consider whether an ensemble of isolated systems has any tendency in the course of time toward a state of statistical equilibrium.

There are certain functions of phase which are constant in time. The distribution of the ensemble with respect to the values of these functions is necessarily invariable, that is, the number of systems within any limits which can be specified in terms of these functions cannot vary in the course of time. The distribution in phase which without violating this condition gives the least value of the average index of probability of phase ($\bar{\eta}$) is unique, and is that in which the

in various ways describe an ensemble in statistical equilibrium, while the same language applied to the critical value of the energy would fail to do so. Thus, if we should say that the ensemble is so distributed that the probability that a system is in any given part of the circle is proportioned to the time which a single system spends in that part, motion in either direction being equally probable, we should perfectly define a distribution in statistical equilibrium for any value of the energy except the critical value mentioned above, but for this value of the energy all the probabilities in question would vanish unless the highest point is included in the part of the circle considered, in which case the probability is unity, or forms one of its limits, in which case the probability is indeterminate. Compare the foot-note on page 118.

A still more simple example is afforded by the uniform motion of a material point in a straight line. Here the impossibility of statistical equilibrium is not limited to any particular energy, and the canonical distribution as well as the microcanonical is impossible.

These examples are mentioned here in order to show the necessity of caution in the application of the above principle, with respect to the question whether we have to do with a true case of statistical equilibrium.

Another point in respect to which caution must be exercised is that the *part* of an ensemble of which the theorem of the return of systems is asserted should be entirely defined by *limits* within which it is contained, and not by any such condition as that a certain function of phase shall have a given value. This is necessary in order that the part of the ensemble which is considered should be any assignable fraction of the whole. Thus, if we have a canonical ensemble consisting of material points in vertical circles, the theorem of the return of systems may be applied to a part of the ensemble defined as contained in a given part of the circle. But it may not be applied in all cases to a part of the ensemble defined as contained in a given part of the circle and having a given energy. It would, in fact, express the exact opposite of the truth when the given energy is the critical value mentioned above.

index of probability (η) is a function of the functions mentioned.* It is therefore a permanent distribution,† and the only permanent distribution consistent with the invariability of the distribution with respect to the functions of phase which are constant in time.

It would seem, therefore, that we might find a sort of measure of the deviation of an ensemble from statistical equilibrium in the excess of the average index above the minimum which is consistent with the condition of the invariability of the distribution with respect to the constant functions of phase. But we have seen that the index of probability is constant in time for each system of the ensemble. The average index is therefore constant, and we find by this method no approach toward statistical equilibrium in the course of time.

Yet we must here exercise great caution. One function may approach indefinitely near to another function, while some quantity determined by the first does not approach the corresponding quantity determined by the second. A line joining two points may approach indefinitely near to the straight line joining them, while its length remains constant. We may find a closer analogy with the case under consideration in the effect of stirring an incompressible liquid.‡ In space of $2n$ dimensions the case might be made analytically identical with that of an ensemble of systems of n degrees of freedom, but the analogy is perfect in ordinary space. Let us suppose the liquid to contain a certain amount of coloring matter which does not affect its hydrodynamic properties. Now the state in which the density of the coloring matter is uniform, *i. e.*, the state of perfect mixture, which is a sort of state of equilibrium in this respect that the distribution of the coloring matter in space is not affected by the internal motions of the liquid, is characterized by a minimum

* See Chapter XI, Theorem IV.

† See Chapter IV, *sub init.*

‡ By *liquid* is here meant the continuous body of theoretical hydrodynamics, and not anything of the molecular structure and molecular motions of real liquids.

value of the average square of the density of the coloring matter. Let us suppose, however, that the coloring matter is distributed with a variable density. If we give the liquid any motion whatever, subject only to the hydrodynamic law of incompressibility, — it may be a steady flux, or it may vary with the time, — the density of the coloring matter at any same point of the liquid will be unchanged, and the average square of this density will therefore be unchanged. Yet no fact is more familiar to us than that stirring tends to bring a liquid to a state of uniform mixture, or uniform densities of its components, which is characterized by minimum values of the average squares of these densities. It is quite true that in the physical experiment the result is hastened by the process of diffusion, but the result is evidently not dependent on that process.

The contradiction is to be traced to the notion of the *density* of the coloring matter, and the process by which this quantity is evaluated. This quantity is the limiting ratio of the quantity of the coloring matter in an element of space to the volume of that element. Now if we should take for our elements of volume, after any amount of stirring, the spaces occupied by the same portions of the liquid which originally occupied any given system of elements of volume, the densities of the coloring matter, thus estimated, would be identical with the original densities as determined by the given system of elements of volume. Moreover, if at the end of any finite amount of stirring we should take our elements of volume in any ordinary form but sufficiently small, the average square of the density of the coloring matter, as determined by such element of volume, would approximate to any required degree to its value before the stirring. But if we take any element of space of fixed position and dimensions, we may continue the stirring so long that the densities of the colored liquid estimated for these fixed elements will approach a uniform limit, viz., that of perfect mixture.

The case is evidently one of those in which the limit of a limit has different values, according to the order in which we apply the processes of taking a limit. If treating the elements of volume as constant, we continue the stirring indefinitely, we get a uniform density, a result not affected by making the elements as small as we choose; but if treating the amount of stirring as finite, we diminish indefinitely the elements of volume, we get exactly the same distribution in density as before the stirring, a result which is not affected by continuing the stirring as long as we choose. The question is largely one of language and definition. One may perhaps be allowed to say that a finite amount of stirring will not affect the mean square of the density of the coloring matter, but an infinite amount of stirring may be regarded as producing a condition in which the mean square of the density has its minimum value, and the density is uniform. We may certainly say that a sensibly uniform density of the colored component may be produced by stirring. Whether the time required for this result would be long or short depends upon the nature of the motion given to the liquid, and the fineness of our method of evaluating the density.

All this may appear more distinctly if we consider a special case of liquid motion. Let us imagine a cylindrical mass of liquid of which one sector of 90° is black and the rest white. Let it have a motion of rotation about the axis of the cylinder in which the angular velocity is a function of the distance from the axis. In the course of time the black and the white parts would become drawn out into thin ribbons, which would be wound spirally about the axis. The thickness of these ribbons would diminish without limit, and the liquid would therefore tend toward a state of perfect mixture of the black and white portions. That is, in any given element of space, the proportion of the black and white would approach 1 : 3 as a limit. Yet after any finite time, the total volume would be divided into two parts, one of which would consist of the white liquid exclusively, and the other of the black exclusively. If the coloring matter, instead of being distributed initially with a uniform density throughout a section of the cylinder, were distributed with a density represented by any arbitrary function of the cylindrical coördinates r, θ and z, the effect of the same motion continued indefinitely would be an approach to a condition in which the density is a function of r and z alone. In this limiting condition, the average square of the density would be less than in the original condition, when the density was supposed to vary with θ, although after any finite time the average square of the density would be the same as at first.

If we limit our attention to the motion in a single plane perpendicular to the axis of the cylinder, we have something which is almost identical with a diagrammatic representation of the changes in distribution in phase of an ensemble of systems of one degree of freedom, in which the motion is periodic, the period varying with the energy, as in the case of a pendulum swinging in a circular arc. If the coördinates and momenta of the systems are represented by rectangular coördinates in the diagram, the points in the diagram representing the changing phases of moving systems, will move about the origin in closed curves of constant energy. The motion will be such that areas bounded by points representing moving systems will be preserved. The only difference between the motion of the liquid and the motion in the diagram is that in one case the paths are circular, and in the other they differ more or less from that form.

When the energy is proportional to $p^2 + q^2$ the curves of constant energy are circles, and the period is independent of the energy. There is then no tendency toward a state of statistical equilibrium. The diagram turns about the origin without change of form. This corresponds to the case of liquid motion, when the liquid revolves with a uniform angular velocity like a rigid solid.

The analogy between the motion of an ensemble of systems in an extension-in-phase and a steady current in an incompressible liquid, and the diagrammatic representation of the case of one degree of freedom, which appeals to our geometrical intuitions, may be sufficient to show how the conservation of density in phase, which involves the conservation of the average value of the index of probability of phase, is consistent with an approach to a limiting condition in which that average value is less. We might perhaps fairly infer from such considerations as have been adduced that an approach to a limiting condition of statistical equilibrium is the general rule, when the initial condition is not of that character. But the subject is of such importance that it seems desirable to give it farther consideration.

Let us suppose that the total extension-in-phase for the kind of system considered to be divided into equal elements (DV) which are very small but not infinitely small. Let us imagine an ensemble of systems distributed in this extension in a manner represented by the index of probability η, which is an arbitrary function of the phase subject only to the restriction expressed by equation (46) of Chapter I. We shall suppose the elements DV to be so small that η may in general be regarded as sensibly constant within any one of them at the initial moment. Let the path of a system be defined as the series of phases through which it passes.

At the initial moment (t') a certain system is in an element of extension DV'. Subsequently, at the time t'', the same system is in the element DV''. Other systems which were at first in DV' will at the time t'' be in DV'', but not all, probably. The systems which were at first in DV' will at the time t'' occupy an extension-in-phase exactly as large as at first. But it will probably be distributed among a very great number of the elements (DV) into which we have divided the total extension-in-phase. If it is not so, we can generally take a later time at which it will be so. There will be exceptions to this for particular laws of motion, but we will confine ourselves to what may fairly be called the general case. Only a very small part of the systems initially in DV' will be found in DV'' at the time t'', and those which are found in DV'' at that time were at the initial moment distributed among a very large number of elements DV.

What is important for our purpose is the value of η, the index of probability of phase in the element DV'' at the time

t''. In the part of DV'' occupied by systems which at the time t' were in DV' the value of η will be the same as its value in DV' at the time t', which we shall call η'. In the parts of DV'' occupied by systems which at t' were in elements very near to DV' we may suppose the value of η to vary little from η'. We cannot assume this in regard to parts of DV'' occupied by systems which at t' were in elements remote from DV'. We want, therefore, some idea of the nature of the extension-in-phase occupied at t' by the systems which at t'' will occupy DV''. Analytically, the problem is identical with finding the extension occupied at t'' by the systems which at t' occupied DV'. Now the systems in DV'' which lie on the same *path* as the system first considered, evidently arrived at DV'' at nearly the same time, and must have left DV' at nearly the same time, and therefore at t' were in or near DV'. We may therefore take η' as the value for these systems. The same essentially is true of systems in DV'' which lie on paths very close to the path already considered. But with respect to paths passing through DV' and DV'', but not so close to the first path, we cannot assume that the time required to pass from DV' to DV'' is nearly the same as for the first path. The difference of the times required may be small in comparison with $t''-t'$, but as this interval can be as large as we choose, the difference of the times required in the different paths has no limit to its possible value. Now if the case were one of statistical equilibrium, the value of η would be constant in any path, and if all the paths which pass through DV'' also pass through or near DV', the value of η throughout DV'' will vary little from η'. But when the case is not one of statistical equilibrium, we cannot draw any such conclusion. The only conclusion which we can draw with respect to the phase at t' of the systems which at t'' are in DV'' is that they are nearly on the same path.

Now if we should make a new estimate of indices of probability of phase at the time t'', using for this purpose the elements DV, — that is, if we should divide the number of systems in DV'', for example, by the total number of systems, and also by the extension-in-phase of the element, and take the logarithm of the quotient, we would get a number which would be less than the average value of η for the systems within DV'' based on the distribution in phase at the time t'.* Hence the average value of η for the whole ensemble of systems based on the distribution at t'' will be less than the average value based on the distribution at t'.

We must not forget that there are exceptions to this general rule. These exceptions are in cases in which the laws of motion are such that systems having small differences of phase will continue always to have small differences of phase.

It is to be observed that if the average index of probability in an ensemble may be said in some sense to have a less value at one time than at another, it is not necessarily priority in time which determines the greater average index. If a distribution, which is not one of statistical equilibrium, should be given for a time t', and the distribution at an earlier time t'' should be defined as that given by the corresponding phases, if we increase the interval leaving t' fixed and taking t'' at an earlier and earlier date, the distribution at t'' will in general approach a limiting distribution which is in statistical equilibrium. The determining difference in such cases is that between a definite distribution at a definite time and the limit of a varying distribution when the moment considered is carried either forward or backward indefinitely.†

But while the distinction of prior and subsequent events may be immaterial with respect to mathematical fictions, it is quite otherwise with respect to the events of the real world. It should not be forgotten, when our ensembles are chosen to illustrate the probabilities of events in the real world, that

* See Chapter XI, Theorem IX.

† One may compare the kinematical truism that when two points are moving with uniform velocities, (with the single exception of the case where the relative motion is zero,) their mutual distance at any definite time is less than for $t = \infty$, or $t = -\infty$.

while the probabilities of subsequent events may often be determined from the probabilities of prior events, it is rarely the case that probabilities of prior events can be determined from those of subsequent events, for we are rarely justified in excluding the consideration of the antecedent probability of the prior events.

It is worthy of notice that to take a system at random from an ensemble at a date chosen at random from several given dates, t', t'', etc., is practically the same thing as to take a system at random from the ensemble composed of all the systems of the given ensemble in their phases at the time t', together with the same systems in their phases at the time t'', etc. By Theorem VIII of Chapter XI this will give an ensemble in which the average index of probability will be less than in the given ensemble, except in the case when the distribution in the given ensemble is the same at the times t', t'', etc. Consequently, any indefiniteness in the time in which we take a system at random from an ensemble has the practical effect of diminishing the average index of the ensemble from which the system may be supposed to be drawn, except when the given ensemble is in statistical equilibrium.

CHAPTER XIII.

EFFECT OF VARIOUS PROCESSES ON AN ENSEMBLE OF SYSTEMS.

In the last chapter and in Chapter I we have considered the changes which take place in the course of time in an ensemble of isolated systems. Let us now proceed to consider the changes which will take place in an ensemble of systems under external influences. These external influences will be of two kinds, the variation of the coördinates which we have called *external*, and the action of other ensembles of systems. The essential difference of the two kinds of influence consists in this, that the bodies to which the external coördinates relate are not distributed in phase, while in the case of interaction of the systems of two ensembles, we have to regard the fact that both are distributed in phase. To find the effect produced on the ensemble with which we are principally concerned, we have therefore to consider single values of what we have called external coördinates, but an infinity of values of the internal coördinates of any other ensemble with which there is interaction.

Or, — to regard the subject from another point of view, — the action between an unspecified system of an ensemble and the bodies represented by the external coördinates, is the action between a system imperfectly determined with respect to phase and one which is perfectly determined; while the interaction between two unspecified systems belonging to different ensembles is the action between two systems both of which are imperfectly determined with respect to phase.*

We shall suppose the ensembles which we consider to be distributed in phase in the manner described in Chapter I, and

* In the development of the subject, we shall find that this distinction corresponds to the distinction in thermodynamics between mechanical and thermal action.

represented by the notations of that chapter, especially by the index of probability of phase (η). There are therefore $2n$ independent variations in the phases which constitute the ensembles considered. This excludes ensembles like the microcanonical, in which, as energy is constant, there are only $2n-1$ independent variations of phase. This seems necessary for the purposes of a general discussion. For although we may imagine a microcanonical ensemble to have a permanent existence when isolated from external influences, the effect of such influences would generally be to destroy the uniformity of energy in the ensemble. Moreover, since the microcanonical ensemble may be regarded as a limiting case of such ensembles as are described in Chapter I, (and that in more than one way, as shown in Chapter X,) the exclusion is rather formal than real, since any properties which belong to the microcanonical ensemble could easily be derived from those of the ensembles of Chapter I, which in a certain sense may be regarded as representing the general case.

Let us first consider the effect of variation of the external coördinates. We have already had occasion to regard these quantities as variable in the differentiation of certain equations relating to ensembles distributed according to certain laws called canonical or microcanonical. That variation of the external coördinates was, however, only carrying the attention of the mind from an ensemble with certain values of the external coördinates, and distributed in phase according to some general law depending upon those values, to another ensemble with different values of the external coördinates, and with the distribution changed to conform to these new values.

What we have now to consider is the effect which would actually result in the course of time in an ensemble of systems in which the external coördinates should be varied in any arbitrary manner. Let us suppose, in the first place, that these coördinates are varied abruptly at a given instant, being constant both before and after that instant. By the definition of the external coördinates it appears that this variation does not affect the phase of any system of the ensemble at the time when it takes place. Therefore it does not affect the index of probability of phase (η) of any system, or the average value of the index ($\bar{\eta}$) at that time. And if these quantities are constant in time before the variation of the external coördinates, and after that variation, their constancy in time is not interrupted by that variation. In fact, in the demonstration of the conservation of probability of phase in Chapter I, the variation of the external coördinates was not excluded.

But a variation of the external coördinates will in general disturb a previously existing state of statistical equilibrium. For, although it does not affect (at the first instant) the distribution-in-phase, it does affect the condition necessary for equilibrium. This condition, as we have seen in Chapter IV, is that the index of probability of phase shall be a function of phase which is constant in time for moving systems. Now a change in the external coördinates, by changing the forces which act on the systems, will change the nature of the functions of phase which are constant in time. Therefore, the distribution in phase which was one of statistical equilibrium for the old values of the external coördinates, will not be such for the new values.

Now we have seen, in the last chapter, that when the distribution-in-phase is not one of statistical equilibrium, an ensemble of systems may, and in general will, after a longer or shorter time, come to a state which may be regarded, if very small differences of phase are neglected, as one of statistical equilibrium, and in which consequently the average value of the index ($\bar{\eta}$) is less than at first. It is evident, therefore, that a variation of the external coördinates, by disturbing a state of statistical equilibrium, may indirectly cause a diminution, (in a certain sense at least,) of the value of $\bar{\eta}$.

But if the change in the external coördinates is very small, the change in the distribution necessary for equilibrium will in general be correspondingly small. Hence, the original distribution in phase, since it differs little from one which would be in statistical equilibrium with the new values of the external coördinates, may be supposed to have a value of $\bar{\eta}$

which differs by a small quantity of the second order from the minimum value which characterizes the state of statistical equilibrium. And the diminution in the average index resulting in the course of time from the very small change in the external coördinates, cannot exceed this small quantity of the second order.

Hence also, if the change in the external coördinates of an ensemble initially in statistical equilibrium consists in successive very small changes separated by very long intervals of time in which the disturbance of statistical equilibrium becomes sensibly effaced, the final diminution in the average index of probability will in general be negligible, although the total change in the external coördinates is large. The result will be the same if the change in the external coördinates takes place continuously but sufficiently slowly.

Even in cases in which there is no tendency toward the restoration of statistical equilibrium in the lapse of time, a variation of external coördinates which would cause, if it took place in a short time, a great disturbance of a previous state of equilibrium, may, if sufficiently distributed in time, produce no sensible disturbance of the statistical equilibrium.

Thus, in the case of three degrees of freedom, let the systems be heavy points suspended by elastic massless cords, and let the ensemble be distributed in phase with a density proportioned to some function of the energy, and therefore in statistical equilibrium. For a change in the external coördinates, we may take a horizontal motion of the point of suspension. If this is moved a given distance, the resulting disturbance of the statistical equilibrium may evidently be diminished indefinitely by diminishing the velocity of the point of suspension. This will be true if the law of elasticity of the string is such that the period of vibration is independent of the energy, in which case there is no tendency in the course of time toward a state of statistical equilibrium, as well as in the more general case, in which there is a tendency toward statistical equilibrium.

That something of this kind will be true in general, the following considerations will tend to show.

We define a path as the series of phases through which a system passes in the course of time when the external coördinates have fixed values. When the external coördinates are varied, paths are changed. The path of a phase is the path to which that phase belongs. With reference to any ensemble of systems we shall denote by $\overline{D}|_p$ the average value of the density-in-phase in a path. This implies that we have a measure for comparing different portions of the path. We shall suppose the time required to traverse any portion of a path to be its measure for the purpose of determining this average.

With this understanding, let us suppose that a certain ensemble is in statistical equilibrium. In every element of extension-in-phase, therefore, the density-in-phase D is equal to its path-average $\overline{D}|_p$. Let a sudden small change be made in the external coördinates. The statistical equilibrium will be disturbed and we shall no longer have $D = \overline{D}|_p$ everywhere. This is not because D is changed, but because $\overline{D}|_p$ is changed, the paths being changed. It is evident that if $D > \overline{D}|_p$ in a part of a path, we shall have $D < \overline{D}|_p$ in other parts of the same path.

Now, if we should imagine a further change in the external coördinates of the same kind, we should expect it to produce an effect of the same kind. But the manner in which the second effect will be superposed on the first will be different, according as it occurs immediately after the first change or after an interval of time. If it occurs immediately after the first change, then in any element of phase in which the first change produced a positive value of $D - \overline{D}|_p$ the second change will add a positive value to the first positive value, and where $D - \overline{D}|_p$ was negative, the second change will add a negative value to the first negative value.

But if we wait a sufficient time before making the second change in the external coördinates, so that systems have passed from elements of phase in which $D - \overline{D}|_p$ was originally positive to elements in which it was originally negative, and *vice versa*, (the systems carrying with them the values of $D - \overline{D}|_p$,) the positive values of $D - \overline{D}|_p$ caused by the second change will be in part superposed on negative values due to the first change, and *vice versa.*

The disturbance of statistical equilibrium, therefore, produced by a given change in the values of the external coordinates may be very much diminished by dividing the change into two parts separated by a sufficient interval of time, and a sufficient interval of time for this purpose is one in which the phases of the individual systems are entirely unlike the first, so that any individual system is differently affected by the change, although the whole ensemble is affected in nearly the same way. Since there is no limit to the diminution of the disturbance of equilibrium by division of the change in the external coördinates, we may suppose as a general rule that by diminishing the velocity of the changes in the external coördinates, a given change may be made to produce a very small disturbance of statistical equilibrium.

If we write $\bar{\eta}'$ for the value of the average index of probability before the variation of the external coördinates, and $\bar{\eta}''$ for the value after this variation, we shall have in any case

$$\bar{\eta}'' \leqq \bar{\eta}'$$

as the simple result of the variation of the external coördinates. This may be compared with the thermodynamic theorem that the entropy of a body cannot be diminished by mechanical (as distinguished from thermal) action.*

If we have (approximate) statistical equilibrium between the times t' and t'' (corresponding to $\bar{\eta}'$ and $\bar{\eta}''$), we shall have approximately

$$\bar{\eta}' = \bar{\eta}'',$$

which may be compared with the thermodynamic theorem that the entropy of a body is not (sensibly) affected by mechanical action, during which the body is at each instant (sensibly) in a state of thermodynamic equilibrium.

Approximate statistical equilibrium may usually be attained

* The correspondences to which the reader's attention is called are between $-\eta$ and entropy, and between Θ and temperature.

by a sufficiently slow variation of the external coördinates, just as approximate thermodynamic equilibrium may usually be attained by sufficient slowness in the mechanical operations to which the body is subject.

We now pass to the consideration of the effect on an ensemble of systems which is produced by the action of other ensembles with which it is brought into dynamical connection. In a previous chapter * we have imagined a dynamical connection arbitrarily created between the systems of two ensembles. We shall now regard the action between the systems of the two ensembles as a result of the variation of the external coördinates, which causes such variations of the internal coördinates as to bring the systems of the two ensembles within the range of each other's action.

Initially, we suppose that we have two separate ensembles of systems, E_1 and E_2. The numbers of degrees of freedom of the systems in the two ensembles will be denoted by n_1 and n_2 respectively, and the probability-coefficients by e^{η_1} and e^{η_2}. Now we may regard any system of the first ensemble combined with any system of the second as forming a single system of $n_1 + n_2$ degrees of freedom. Let us consider the ensemble (E_{12}) obtained by thus combining each system of the first ensemble with each of the second.

At the initial moment, which may be specified by a single accent, the probability-coefficient of any phase of the combined systems is evidently the product of the probability-coefficients of the phases of which it is made up. This may be expressed by the equation,

$$e^{\eta_{12}'} = e^{\eta_1'} e^{\eta_2'}, \tag{455}$$

or

$$\eta_{12}' = \eta_1' + \eta_2', \tag{456}$$

which gives

$$\bar{\eta}_{12}' = \bar{\eta}_1' + \bar{\eta}_2'. \tag{457}$$

The forces tending to vary the internal coördinates of the combined systems, together with those exerted by either system upon the bodies represented by the coördinates called

* See Chapter IV, page 37.

external, may be derived from a single force-function, which, taken negatively, we shall call the potential energy of the combined systems and denote by ϵ_{12}. But we suppose that initially none of the systems of the two ensembles E_1 and E_2 come within range of each other's action, so that the potential energy of the combined system falls into two parts relating separately to the systems which are combined. The same is obviously true of the kinetic energy of the combined compound system, and therefore of its total energy. This may be expressed by the equation

$$\epsilon_{12}' = \epsilon_1' + \epsilon_2', \tag{458}$$

which gives

$$\bar{\epsilon}_{12}' = \bar{\epsilon}_1' + \bar{\epsilon}_2'. \tag{459}$$

Let us now suppose that in the course of time, owing to the motion of the bodies represented by the coördinates called external, the forces acting on the systems and consequently their positions are so altered, that the systems of the ensembles E_1 and E_2 are brought within range of each other's action, and after such mutual influence has lasted for a time, by a further change in the external coördinates, perhaps a return to their original values, the systems of the two original ensembles are brought again out of range of each other's action. Finally, then, at a time specified by double accents, we shall have as at first

$$\bar{\epsilon}_{12}'' = \bar{\epsilon}_1'' + \bar{\epsilon}_2''. \tag{460}$$

But for the indices of probability we must write *

$$\bar{\eta}_1'' + \bar{\eta}_2'' \leqq \bar{\eta}_{12}''. \tag{461}$$

The considerations adduced in the last chapter show that it is safe to write

$$\bar{\eta}_{12}'' \leqq \bar{\eta}_{12}'. \tag{462}$$

We have therefore

$$\bar{\eta}_1'' + \bar{\eta}_2'' \leqq \bar{\eta}_1' + \bar{\eta}_2', \tag{463}$$

which may be compared with the thermodynamic theorem that

* See Chapter XI, Theorem VII.

the thermal contact of two bodies may increase but cannot diminish the sum of their entropies.

Let us especially consider the case in which the two original ensembles were both canonically distributed in phase with the respective moduli Θ_1 and Θ_2. We have then, by Theorem III of Chapter XI,

$$\bar{\eta}_1' + \frac{\bar{\epsilon}_1'}{\Theta_1} \leqq \bar{\eta}_1'' + \frac{\bar{\epsilon}_1''}{\Theta_1} \tag{464}$$

$$\bar{\eta}_2' + \frac{\bar{\epsilon}_2'}{\Theta_2} \leqq \bar{\eta}_2'' + \frac{\bar{\epsilon}_2''}{\Theta_2} \tag{465}$$

Whence with (463) we have

$$\frac{\bar{\epsilon}_1'}{\Theta_1} + \frac{\bar{\epsilon}_2'}{\Theta_2} \leqq \frac{\bar{\epsilon}_1''}{\Theta_1} + \frac{\bar{\epsilon}_2''}{\Theta_2} \tag{466}$$

or
$$\frac{\bar{\epsilon}_1'' - \bar{\epsilon}_1'}{\Theta_1} + \frac{\bar{\epsilon}_2'' - \bar{\epsilon}_2'}{\Theta_2} \geqq 0. \tag{467}$$

If we write $\overline{W}$ for the average work done by the combined systems on the external bodies, we have by the principle of the conservation of energy

$$\overline{W} = \bar{\epsilon}_{12}' - \bar{\epsilon}_{12}'' = \bar{\epsilon}_1' - \bar{\epsilon}_1'' + \bar{\epsilon}_2' - \bar{\epsilon}_2''. \tag{468}$$

Now if $\overline{W}$ is negligible, we have

$$\bar{\epsilon}_1'' - \bar{\epsilon}_1' = -(\bar{\epsilon}_2'' - \bar{\epsilon}_2') \tag{469}$$

and (467) shows that the ensemble which has the greater modulus must lose energy. This result may be compared to the thermodynamic principle, that when two bodies of different temperatures are brought together, that which has the higher temperature will lose energy.

Let us next suppose that the ensemble E_2 is originally canonically distributed with the modulus Θ_2, but leave the distribution of the other arbitrary. We have, to determine the result of a similar process,

$$\bar{\eta}_1'' + \bar{\eta}_2'' \leqq \bar{\eta}_1' + \bar{\eta}_2'$$

$$\bar{\eta}_2' + \frac{\bar{\epsilon}_2'}{\Theta_2} \leqq \bar{\eta}_2'' + \frac{\bar{\epsilon}_2''}{\Theta_2}$$

Hence
$$\bar{\eta}_1'' + \frac{\bar{\epsilon}_2'}{\Theta_2} \leqq \bar{\eta}_1' + \frac{\bar{\epsilon}_2''}{\Theta_2} \tag{470}$$

which may be written

$$\bar{\eta}_1' - \bar{\eta}_1'' \geqq \frac{\bar{\epsilon}_2' - \bar{\epsilon}_2''}{\Theta_2} \tag{471}$$

This may be compared with the thermodynamic principle that when a body (which need not be in thermal equilibrium) is brought into thermal contact with another of a given temperature, the increase of entropy of the first cannot be less (algebraically) than the loss of heat by the second divided by its temperature. Where $\overline{W}$ is negligible, we may write

$$\bar{\eta}_1'' + \frac{\bar{\epsilon}_1''}{\Theta_2} \leqq \bar{\eta}_1' + \frac{\bar{\epsilon}_1'}{\Theta_2} \tag{472}$$

Now, by Theorem III of Chapter XI, the quantity

$$\bar{\eta}_1 + \frac{\bar{\epsilon}_1}{\Theta_2} \tag{473}$$

has a minimum value when the ensemble to which $\bar{\eta}_1$ and $\bar{\epsilon}_1$ relate is distributed canonically with the modulus Θ_2. If the ensemble had originally this distribution, the sign $<$ in (472) would be impossible. In fact, in this case, it would be easy to show that the preceding formulae on which (472) is founded would all have the sign $=$. But when the two ensembles are not both originally distributed canonically with the same modulus, the formulae indicate that the quantity (473) may be diminished by bringing the ensemble to which ϵ_1 and η_1 relate into connection with another which is canonically distributed with modulus Θ_2, and therefore, that by repeated operations of this kind the ensemble of which the original distribution was entirely arbitrary might be brought approximately into a state of canonical distribution with the modulus Θ_2. We may compare this with the thermodynamic principle that a body of which the original thermal state may be entirely arbitrary, may be brought approximately into a state of thermal equilibrium with any given temperature by repeated connections with other bodies of that temperature.

11

Let us now suppose that we have a certain number of ensembles, E_0, E_1, E_2, etc., distributed canonically with the respective moduli Θ_0, Θ_1, Θ_2, etc. By variation of the external coördinates of the ensemble E_0, let it be brought into connection with E_1, and then let the connection be broken. Let it then be brought into connection with E_2, and then let that connection be broken. Let this process be continued with respect to the remaining ensembles. We do not make the assumption, as in some cases before, that the work connected with the variation of the external coördinates is a negligible quantity. On the contrary, we wish especially to consider the case in which it is large. In the final state of the ensemble E_0, let us suppose that the external coördinates have been brought back to their original values, and that the average energy $(\bar{\epsilon}_0)$ is the same as at first.

In our usual notations, using one and two accents to distinguish original and final values, we get by repeated applications of the principle expressed in (463)

$$\bar{\eta}_0' + \bar{\eta}_1' + \bar{\eta}_2' + \text{etc.} \geqq \bar{\eta}_0'' + \bar{\eta}_1'' + \bar{\eta}_2'' + \text{etc.} \tag{474}$$

But by Theorem III of Chapter XI,

$$\bar{\eta}_0'' + \frac{\bar{\epsilon}_0''}{\Theta_0} \geqq \bar{\eta}_0' + \frac{\bar{\epsilon}_0'}{\Theta_0}, \tag{475}$$

$$\bar{\eta}_1'' + \frac{\bar{\epsilon}_1''}{\Theta_1} \geqq \bar{\eta}_1' + \frac{\bar{\epsilon}_1'}{\Theta_1}, \tag{476}$$

$$\bar{\eta}_2'' + \frac{\bar{\epsilon}_2''}{\Theta_2} \geqq \bar{\eta}_2' + \frac{\bar{\epsilon}_2'}{\Theta_2}, \tag{477}$$

etc.

Hence
$$\frac{\bar{\epsilon}_0''}{\Theta_0} + \frac{\bar{\epsilon}_1''}{\Theta_1} + \frac{\bar{\epsilon}_2''}{\Theta_2} + \text{etc.} \geqq \frac{\bar{\epsilon}_0'}{\Theta_0} + \frac{\bar{\epsilon}_1'}{\Theta_1} + \frac{\bar{\epsilon}_2'}{\Theta_2} + \text{etc.} \tag{478}$$

or, since
$$\bar{\epsilon}_0' = \bar{\epsilon}_0'',$$

$$0 \geqq \frac{\bar{\epsilon}_1' - \bar{\epsilon}_1''}{\Theta_1} + \frac{\bar{\epsilon}_2' - \bar{\epsilon}_2''}{\Theta_2} + \text{etc.} \tag{479}$$

If we write $\overline{W}$ for the average work done on the bodies represented by the external coördinates, we have

$$\bar{\epsilon}_1' - \bar{\epsilon}_1'' + \bar{\epsilon}_2' - \bar{\epsilon}_2'' + \text{etc.} = \overline{W}. \tag{480}$$

If E_0, E_1, and E_2 are the only ensembles, we have

$$W \leqq \frac{\Theta_1 - \Theta_2}{\Theta_1} (\bar{\epsilon}_1' - \bar{\epsilon}_1''). \tag{481}$$

It will be observed that the relations expressed in the last three formulae between $\overline{W}$, $\bar{\epsilon}_1 - \bar{\epsilon}_1''$, $\bar{\epsilon}_2' - \bar{\epsilon}_2''$, etc., and Θ_1, Θ_2, etc. are precisely those which hold in a Carnot's cycle for the work obtained, the energy lost by the several bodies which serve as heaters or coolers, and their initial temperatures.

It will not escape the reader's notice, that while from one point of view the operations which are here described are quite beyond our powers of actual performance, on account of the impossibility of handling the immense number of systems which are involved, yet from another point of view the operations described are the most simple and accurate means of representing what actually takes place in our simplest experiments in thermodynamics. The states of the bodies which we handle are certainly not known to us exactly. What we know about a body can generally be described most accurately and most simply by saying that it is one taken at random from a great number (ensemble) of bodies which are completely described. If we bring it into connection with another body concerning which we have a similar limited knowledge, the state of the two bodies is properly described as that of a pair of bodies taken from a great number (ensemble) of pairs which are formed by combining each body of the first ensemble with each of the second.

Again, when we bring one body into thermal contact with another, for example, in a Carnot's cycle, when we bring a mass of fluid into thermal contact with some other body from which we wish it to receive heat, we may do it by moving the vessel containing the fluid. This motion is mathematically expressed by the variation of the coördinates which determine the position of the vessel. We allow ourselves for the purposes of a theoretical discussion to suppose that the walls of this vessel are incapable of absorbing heat from the fluid.

Yet while we exclude the kind of action which we call thermal between the fluid and the containing vessel, we allow the kind which we call work in the narrower sense, which takes place when the volume of the fluid is changed by the motion of a piston. This agrees with what we have supposed in regard to the external coördinates, which we may vary in any arbitrary manner, and are in this entirely unlike the coordinates of the second ensemble with which we bring the first into connection.

When heat passes in any thermodynamic experiment between the fluid principally considered and some other body, it is actually absorbed and given out by the walls of the vessel, which will retain a varying quantity. This is, however, a disturbing circumstance, which we suppose in some way made negligible, and actually neglect in a theoretical discussion. In our case, we suppose the walls incapable of absorbing energy, except through the motion of the external coördinates, but that they allow the systems which they contain to act directly on one another. Properties of this kind are mathematically expressed by supposing that in the vicinity of a certain surface, the position of which is determined by certain (external) coördinates, particles belonging to the system in question experience a repulsion from the surface increasing so rapidly with nearness to the surface that an infinite expenditure of energy would be required to carry them through it. It is evident that two systems might be separated by a surface or surfaces exerting the proper forces, and yet approach each other closely enough to exert mechanical action on each other.

CHAPTER XIV.

DISCUSSION OF THERMODYNAMIC ANALOGIES.

If we wish to find in rational mechanics an *a priori* foundation for the principles of thermodynamics, we must seek mechanical definitions of temperature and entropy. The quantities thus defined must satisfy (under conditions and with limitations which again must be specified in the language of mechanics) the differential equation

$$d\epsilon = T d\eta - A_1 \, da_1 - A_2 \, da_2 - \text{etc.}, \tag{482}$$

where ϵ, T, and η denote the energy, temperature, and entropy of the system considered, and $A_1 da_1$, etc., the mechanical work (in the narrower sense in which the term is used in thermodynamics, *i. e.*, with exclusion of thermal action) done upon external bodies.

This implies that we are able to distinguish in mechanical terms the thermal action of one system on another from that which we call mechanical in the narrower sense, if not indeed in every case in which the two may be combined, at least so as to specify cases of thermal action and cases of mechanical action.

Such a differential equation moreover implies a finite equation between ϵ, η, and a_1, a_2, etc., which may be regarded as fundamental in regard to those properties of the system which we call thermodynamic, or which may be called so from analogy. This fundamental thermodynamic equation is determined by the fundamental mechanical equation which expresses the energy of the system as function of its momenta and coördinates with those external coördinates (a_1, a_2, etc.) which appear in the differential expression of the work done on external bodies. We have to show the mathematical operations by which the fundamental thermodynamic equation, which in general is an equation of few variables, is derived from the fundamental mechanical equation, which in the case of the bodies of nature is one of an enormous number of variables.

We have also to enunciate in mechanical terms, and to prove, what we call the tendency of heat to pass from a system of higher temperature to one of lower, and to show that this tendency vanishes with respect to systems of the same temperature.

At least, we have to show by *a priori* reasoning that for such systems as the material bodies which nature presents to us, these relations hold with such approximation that they are sensibly true for human faculties of observation. This indeed is all that is really necessary to establish the science of thermodynamics on an *a priori* basis. Yet we will naturally desire to find the exact expression of those principles of which the laws of thermodynamics are the approximate expression. A very little study of the statistical properties of conservative systems of a finite number of degrees of freedom is sufficient to make it appear, more or less distinctly, that the general laws of thermodynamics are the limit toward which the exact laws of such systems approximate, when their number of degrees of freedom is indefinitely increased. And the problem of finding the exact relations, as distinguished from the approximate, for systems of a great number of degrees of freedom, is practically the same as that of finding the relations which hold for any number of degrees of freedom, as distinguished from those which have been established on an empirical basis for systems of a great number of degrees of freedom.

The enunciation and proof of these exact laws, for systems of any finite number of degrees of freedom, has been a principal object of the preceding discussion. But it should be distinctly stated that, if the results obtained when the numbers of degrees of freedom are enormous coincide sensibly with the general laws of thermodynamics, however interesting and significant this coincidence may be, we are still far from having explained the phenomena of nature with respect to these laws. For, as compared with the case of nature, the systems which we have considered are of an ideal simplicity. Although our only assumption is that we are considering conservative systems of a finite number of degrees of freedom, it would seem that this is assuming far too much, so far as the bodies of nature are concerned. The phenomena of radiant heat, which certainly should not be neglected in any complete system of thermodynamics, and the electrical phenomena associated with the combination of atoms, seem to show that the hypothesis of systems of a finite number of degrees of freedom is inadequate for the explanation of the properties of bodies.

Nor do the results of such assumptions in every detail appear to agree with experience. We should expect, for example, that a diatomic gas, so far as it could be treated independently of the phenomena of radiation, or of any sort of electrical manifestations, would have six degrees of freedom for each molecule. But the behavior of such a gas seems to indicate not more than five.

But although these difficulties, long recognized by physicists,* seem to prevent, in the present state of science, any satisfactory explanation of the phenomena of thermodynamics as presented to us in nature, the ideal case of systems of a finite number of degrees of freedom remains as a subject which is certainly not devoid of a theoretical interest, and which may serve to point the way to the solution of the far more difficult problems presented to us by nature. And if the study of the statistical properties of such systems gives us an exact expression of laws which in the limiting case take the form of the received laws of thermodynamics, its interest is so much the greater.

Now we have defined what we have called the *modulus* (Θ) of an ensemble of systems canonically distributed in phase, and what we have called the index of probability (η) of any phase in such an ensemble. It has been shown that between

* See Boltzmann, Sitzb. der Wiener Akad., Bd. LXIII., S. 418, (1871).

the modulus (Θ), the external coördinates (a_1, etc.), and the average values in the ensemble of the energy (ϵ), the index of probability (η), and the external forces (A_1, etc.) exerted by the systems, the following differential equation will hold:

$$d\bar{\epsilon} = -\Theta\, d\bar{\eta} - \bar{A}_1\, da_1 - \bar{A}_2\, da_2 - \text{etc.} \tag{483}$$

This equation, if we neglect the sign of averages, is identical in form with the thermodynamic equation (482), the modulus (Θ) corresponding to temperature, and the index of probability of phase with its sign reversed corresponding to entropy.*

We have also shown that the average square of the anomalies of ϵ, that is, of the deviations of the individual values from the average, is in general of the same order of magnitude as the reciprocal of the number of degrees of freedom, and therefore to human observation the individual values are indistinguishable from the average values when the number of degrees of freedom is very great.† In this case also the anomalies of η are practically insensible. The same is true of the anomalies of the external forces (A_1, etc.), so far as these are the result of the anomalies of energy, so that when these forces are sensibly determined by the energy and the external coördinates, and the number of degrees of freedom is very great, the anomalies of these forces are insensible.

The mathematical operations by which the finite equation between $\bar{\epsilon}$, $\bar{\eta}$, and a_1, etc., is deduced from that which gives the energy (ϵ) of a system in terms of the momenta ($p_1 \ldots p_n$) and coördinates both internal ($q_1 \ldots q_n$) and external (a_1, etc.), are indicated by the equation

$$e^{-\frac{\psi}{\Theta}} = \underset{\text{phases}}{\overset{\text{all}}{\int \ldots \int}} e^{-\frac{\epsilon}{\Theta}} dq_1 \ldots dq_n dp_1 \ldots dp_n, \tag{484}$$

where $$\psi = \Theta\bar{\eta} + \bar{\epsilon}.$$

We have also shown that when systems of different ensembles are brought into conditions analogous to thermal contact, the average result is a passage of energy from the ensemble

* See Chapter IV, pages 44, 45. † See Chapter VII, pages 73–75.

of the greater modulus to that of the less,* or in case of equal moduli, that we have a condition of statistical equilibrium in regard to the distribution of energy.†

Propositions have also been demonstrated analogous to those in thermodynamics relating to a Carnot's cycle,‡ or to the tendency of entropy to increase,§ especially when bodies of different temperature are brought into contact.‖

We have thus precisely defined quantities, and rigorously demonstrated propositions, which hold for any number of degrees of freedom, and which, when the number of degrees of freedom (n) is enormously great, would appear to human faculties as the quantities and propositions of empirical thermodynamics.

It is evident, however, that there may be more than one quantity defined for finite values of n, which approach the same limit, when n is increased indefinitely, and more than one proposition relating to finite values of n, which approach the same limiting form for $n = \infty$. There may be therefore, and there are, other quantities which may be thought to have some claim to be regarded as temperature and entropy with respect to systems of a finite number of degrees of freedom.

The definitions and propositions which we have been considering relate essentially to what we have called a canonical ensemble of systems. This may appear a less natural and simple conception than what we have called a microcanonical ensemble of systems, in which all have the same energy, and which in many cases represents simply the *time-ensemble*, or ensemble of phases through which a single system passes in the course of time.

It may therefore seem desirable to find definitions and propositions relating to these microcanonical ensembles, which shall correspond to what in thermodynamics are based on experience. Now the differential equation

$$d\epsilon = e^{-\phi} V\, d \log V - \overline{A_1}|_\epsilon \, da_1 - \overline{A_2}|_\epsilon \, da_2 - \text{etc.}, \qquad (485)$$

* See Chapter XIII, page 160.

† See Chapter IV, pages 35–37.

‡ See Chapter XIII, pages 162, 163.

§ See Chapter XII, pages 143–151.

‖ See Chapter XIII, page 159.

which has been demonstrated in Chapter X, and which relates to a microcanonical ensemble, $\overline{A_1}|_\epsilon$ denoting the average value of A_1 in such an ensemble, corresponds precisely to the thermodynamic equation, except for the sign of average applied to the external forces. But as these forces are not entirely determined by the energy with the external coördinates, the use of average values is entirely germane to the subject, and affords the readiest means of getting perfectly determined quantities. These averages, which are taken for a microcanonical ensemble, may seem from some points of view a more simple and natural conception than those which relate to a canonical ensemble. Moreover, the energy, and the quantity corresponding to entropy, are free from the sign of average in this equation.

The quantity in the equation which corresponds to entropy is $\log V$, the quantity V being defined as the extension-in-phase within which the energy is less than a certain limiting value (ϵ). This is certainly a more simple conception than the average value in a canonical ensemble of the index of probability of phase. Log V has the property that when it is constant

$$d\epsilon = -\overline{A_1}|_\epsilon\, da_1 - \overline{A_2}|_\epsilon\, da_2 + \text{etc.}, \tag{486}$$

which closely corresponds to the thermodynamic property of entropy, that when it is constant

$$d\epsilon = -A_1\, da_1 - A_2\, da_2 + \text{etc.} \tag{487}$$

The quantity in the equation which corresponds to temperature is $e^{-\phi} V$, or $d\epsilon/d \log V$. In a canonical ensemble, the average value of this quantity is equal to the modulus, as has been shown by different methods in Chapters IX and X.

In Chapter X it has also been shown that if the systems of a microcanonical ensemble consist of parts with separate energies, the average value of $e^{-\phi} V$ for any part is equal to its average value for any other part, and to the uniform value of the same expression for the whole ensemble. This corresponds to the theorem in the theory of heat that in case of thermal equilibrium the temperatures of the parts of a body are equal to one another and to that of the whole body.

Since the energies of the parts of a body cannot be supposed to remain absolutely constant, even where this is the case with respect to the whole body, it is evident that if we regard the temperature as a function of the energy, the taking of average or of probable values, or some other statistical process, must be used with reference to the parts, in order to get a perfectly definite value corresponding to the notion of temperature.

It is worthy of notice in this connection that the average value of the kinetic energy, either in a microcanonical ensemble, or in a canonical, divided by one half the number of degrees of freedom, is equal to $\epsilon^{-\phi} V$, or to its average value, and that this is true not only of the whole system which is distributed either microcanonically or canonically, but also of any part, although the corresponding theorem relating to temperature hardly belongs to empirical thermodynamics, since neither the (inner) kinetic energy of a body, nor its number of degrees of freedom is immediately cognizable to our faculties, and we meet the gravest difficulties when we endeavor to apply the theorem to the theory of gases, except in the simplest case, that of the gases known as monatomic.

But the correspondence between $\epsilon^{-\phi} V$ or $d\epsilon / d \log V$ and temperature is imperfect. If two isolated systems have such energies that

$$\frac{d\epsilon_1}{d \log V_1} = \frac{d\epsilon_2}{d \log V_2},$$

and the two systems are regarded as combined to form a third system with energy

$$\epsilon_{12} = \epsilon_1 + \epsilon_2,$$

we shall not have in general

$$\frac{d\epsilon_{12}}{d \log V_{12}} = \frac{d\epsilon_1}{d \log V_1} = \frac{d\epsilon_2}{d \log V_2},$$

as analogy with temperature would require. In fact, we have seen that

$$\frac{d\epsilon_{12}}{d \log V_{12}} = \overline{\frac{d\epsilon_1}{d \log V_1}}\bigg|_{\epsilon_{12}} = \overline{\frac{d\epsilon_2}{d \log V_2}}\bigg|_{\epsilon_{12}},$$

where the second and third members of the equation denote average values in an ensemble in which the compound system is microcanonically distributed in phase. Let us suppose the two original systems to be identical in nature. Then

$$\epsilon_1 = \epsilon_2 = \overline{\epsilon_1}|_{\epsilon_{12}} = \overline{\epsilon_2}|_{\epsilon_{12}}.$$

The equation in question would require that

$$\frac{d\epsilon_1}{d \log V_1} = \overline{\frac{d\epsilon_1}{d \log V_1}}\Bigg|_{\epsilon_{12}},$$

i. e., that we get the same result, whether we take the value of $d\epsilon_1/d \log V_1$ determined for the average value of ϵ_1 in the ensemble, or take the average value of $d\epsilon_1/d \log V_1$. This will be the case where $d\epsilon_1/d \log V_1$ is a linear function of ϵ_1. Evidently this does not constitute the most general case. Therefore the equation in question cannot be true in general. It is true, however, in some very important particular cases, as when the energy is a quadratic function of the p's and q's, or of the p's alone.* When the equation holds, the case is analogous to that of bodies in thermodynamics for which the specific heat for constant volume is constant.

Another quantity which is closely related to temperature is $d\phi/d\epsilon$. It has been shown in Chapter IX that in a canonical ensemble, if $n > 2$, the average value of $d\phi/d\epsilon$ is $1/\Theta$, and that the most common value of the energy in the ensemble is that for which $d\phi/d\epsilon = 1/\Theta$. The first of these properties may be compared with that of $d\epsilon/d \log V$, which has been seen to have the average value Θ in a canonical ensemble, without restriction in regard to the number of degrees of freedom.

With respect to microcanonical ensembles also, $d\phi/d\epsilon$ has a property similar to what has been mentioned with respect to $d\epsilon/d \log V$. That is, if a system microcanonically distributed in phase consists of two parts with separate energies, and each

* This last case is important on account of its relation to the theory of gases, although it must in strictness be regarded as a limit of possible cases, rather than as a case which is itself possible.

with more than two degrees of freedom, the average values in the ensemble of $d\phi/d\epsilon$ for the two parts are equal to one another and to the value of same expression for the whole. In our usual notations

$$\overline{\frac{d\phi_1}{d\epsilon_1}}\bigg|_{\epsilon_{12}} = \overline{\frac{d\phi_2}{d\epsilon_2}}\bigg|_{\epsilon_{12}} = \frac{d\phi_{12}}{d\epsilon_{12}}$$

if $n_1 > 2$, and $n_2 > 2$.

This analogy with temperature has the same incompleteness which was noticed with respect to $d\epsilon/d\log V$, viz., if two systems have such energies (ϵ_1 and ϵ_2) that

$$\frac{d\phi_1}{d\epsilon_1} = \frac{d\phi_2}{d\epsilon_2},$$

and they are combined to form a third system with energy

$$\epsilon_{12} = \epsilon_1 + \epsilon_2,$$

we shall not have in general

$$\frac{d\phi_{12}}{d\epsilon_{12}} = \frac{d\phi_1}{d\epsilon_1} = \frac{d\phi_2}{d\epsilon_2}.$$

Thus, if the energy is a quadratic function of the p's and q's, we have *

$$\frac{d\phi_1}{d\epsilon_1} = \frac{n_1 - 1}{\epsilon_1}, \qquad \frac{d\phi_2}{d\epsilon_2} = \frac{n_2 - 1}{\epsilon_2},$$

$$\frac{d\phi_{12}}{d\epsilon_{12}} = \frac{n_{12} - 1}{\epsilon_{12}} = \frac{n_1 + n_2 - 1}{\epsilon_1 + \epsilon_2},$$

where n_1, n_2, n_{12}, are the numbers of degrees of freedom of the separate and combined systems. But

$$\frac{d\phi_1}{d\epsilon_1} = \frac{d\phi_2}{d\epsilon_2} = \frac{n_1 + n_2 - 2}{\epsilon_1 + \epsilon_2}.$$

If the energy is a quadratic function of the p's alone, the case would be the same except that we should have $\frac{1}{2}n_1$, $\frac{1}{2}n_2$, $\frac{1}{2}n_{12}$, instead of n_1, n_2, n_{12}. In these particular cases, the analogy

* See foot-note on page 93. We have here made the least value of the energy consistent with the values of the external coördinates zero instead of ϵ_a, as is evidently allowable when the external coördinates are supposed invariable.

between $d\epsilon/d\log V$ and temperature would be complete, as has already been remarked. We should have

$$\frac{d\epsilon_1}{d\log V_1}=\frac{\epsilon_1}{n_1},\qquad \frac{d\epsilon_2}{d\log V_2}=\frac{\epsilon_2}{n_2},$$

$$\frac{d\epsilon_{12}}{d\log V_{12}}=\frac{\epsilon_{12}}{n_{12}}=\frac{d\epsilon_1}{d\log V_1}=\frac{d\epsilon_2}{d\log V_2},$$

when the energy is a quadratic function of the p's and q's, and similar equations with $\frac{1}{2}n_1$, $\frac{1}{2}n_2$, $\frac{1}{2}n_{12}$, instead of n_1, n_2, n_{12}, when the energy is a quadratic function of the p's alone.

More characteristic of $d\phi/d\epsilon$ are its properties relating to most probable values of energy. If a system having two parts with separate energies and each with more than two degrees of freedom is microcanonically distributed in phase, the most probable division of energy between the parts, in a system taken at random from the ensemble, satisfies the equation

$$\frac{d\phi_1}{d\epsilon_1}=\frac{d\phi_2}{d\epsilon_2}, \tag{488}$$

which corresponds to the thermodynamic theorem that the distribution of energy between the parts of a system, in case of thermal equilibrium, is such that the temperatures of the parts are equal.

To prove the theorem, we observe that the fractional part of the whole number of systems which have the energy of one part (ϵ_1) between the limits ϵ_1' and ϵ_1'' is expressed by

$$e^{-\phi_{12}}\int_{\epsilon_1'}^{\epsilon_1''} e^{\phi_1+\phi_2}\,d\epsilon_1,$$

where the variables are connected by the equation

$$\epsilon_1+\epsilon_2=\text{constant}=\epsilon_{12}.$$

The greatest value of this expression, for a constant infinitesimal value of the difference $\epsilon_1''-\epsilon_1'$, determines a value of ϵ_1, which we may call its most probable value. This depends on the greatest possible value of $\phi_1+\phi_2$. Now if $n_1>2$, and $n_2>2$, we shall have $\phi_1=-\infty$ for the least possible value of

ϵ_1, and $\phi_2 = -\infty$ for the least possible value of ϵ_2. Between these limits ϕ_1 and ϕ_2 will be finite and continuous. Hence $\phi_1 + \phi_2$ will have a maximum satisfying the equation (488).

But if $n_1 \leqq 2$, or $n_2 \leqq 2$, $d\phi_1/d\epsilon_1$ or $d\phi_2/d\epsilon_2$ may be negative, or zero, for all values of ϵ_1 or ϵ_2, and can hardly be regarded as having properties analogous to temperature.

It is also worthy of notice that if a system which is microcanonically distributed in phase has three parts with separate energies, and each with more than two degrees of freedom, the most probable division of energy between these parts satisfies the equation

$$\frac{d\phi_1}{d\epsilon_1} = \frac{d\phi_2}{d\epsilon_2} = \frac{d\phi_3}{d\epsilon_3}.$$

That is, this equation gives the most probable set of values of ϵ_1, ϵ_2, and ϵ_3. But it does not give the most probable value of ϵ_1, or of ϵ_2, or of ϵ_3. Thus, if the energies are quadratic functions of the p's and q's, the most probable division of energy is given by the equation

$$\frac{n_1 - 1}{\epsilon_1} = \frac{n_2 - 1}{\epsilon_1} = \frac{n_3 - 1}{\epsilon_3}.$$

But the most probable value of ϵ_1 is given by

$$\frac{n_1 - 1}{\epsilon_1} = \frac{n_2 + n_3 - 1}{\epsilon_2 + \epsilon_3},$$

while the preceding equations give

$$\frac{n_1 - 1}{\epsilon_1} = \frac{n_2 + n_3 - 2}{\epsilon_2 + \epsilon_3}.$$

These distinctions vanish for very great values of n_1, n_2, n_3. For small values of these numbers, they are important. Such facts seem to indicate that the consideration of the most probable division of energy among the parts of a system does not afford a convenient foundation for the study of thermodynamic analogies in the case of systems of a small number of degrees of freedom. The fact that a certain division of energy is the most probable has really no especial physical importance, except when the ensemble of possible divisions are grouped so closely together that the most probable division may fairly represent the whole. This is in general the case, to a very close approximation, when n is enormously great; it entirely fails when n is small.

If we regard $d\phi/d\epsilon$ as corresponding to the reciprocal of temperature, or, in other words, $d\epsilon/d\phi$ as corresponding to temperature, ϕ will correspond to entropy. It has been defined as $\log(dV/d\epsilon)$. In the considerations on which its definition is founded, it is therefore very similar to $\log V$. We have seen that $d\phi/d\log V$ approaches the value unity when n is very great.*

To form a differential equation on the model of the thermodynamic equation (482), in which $d\epsilon/d\phi$ shall take the place of temperature, and ϕ of entropy, we may write

$$d\epsilon = \left(\frac{d\epsilon}{d\phi}\right)_a d\phi + \left(\frac{d\epsilon}{da_1}\right)_{\phi,a} da_1 + \left(\frac{d\epsilon}{da_2}\right)_{\phi,a} da_2 + \text{etc.,} \tag{489}$$

or

$$d\phi = \frac{d\phi}{d\epsilon} d\epsilon + \frac{d\phi}{da_1} da_1 + \frac{d\phi}{da_2} da_2 + \text{etc.} \tag{490}$$

With respect to the differential coefficients in the last equation, which corresponds exactly to (482) solved with respect to $d\eta$, we have seen that their average values in a canonical ensemble are equal to $1/\Theta$, and the averages of A_1/Θ, A_2/Θ, etc.† We have also seen that $d\epsilon/d\phi$ (or $d\phi/d\epsilon$) has relations to the most probable values of energy in parts of a microcanonical ensemble. That $(d\epsilon/da_1)_{\phi,a}$, etc., have properties somewhat analogous, may be shown as follows.

In a physical experiment, we measure a force by balancing it against another. If we should ask what force applied to increase or diminish a_1 would balance the action of the systems, it would be one which varies with the different systems. But we may ask what single force will make a given value of a_1 the most probable, and we shall find that under certain conditions $(d\epsilon/da_1)_{\phi,a}$ represents that force.

* See Chapter X, pages 120, 121.

† See Chapter IX, equations (321), (327).

To make the problem definite, let us consider a system consisting of the original system together with another having the coördinates a_1, a_2, etc., and forces A_1', A_2', etc., tending to increase those coördinates. These are in addition to the forces A_1, A_2, etc., exerted by the original system, and are derived from a force-function $(-\epsilon_q')$ by the equations

$$A_1' = -\frac{d\epsilon_q'}{da_1}, \qquad A_2' = -\frac{d\epsilon_q'}{da_2}, \qquad \text{etc.}$$

For the energy of the whole system we may write

$$\mathrm{E} = \epsilon + \epsilon_q' + \tfrac{1}{2} m_1 \dot{a}_1^2 + \tfrac{1}{2} m_2 \dot{a}_2^2 + \text{etc.},$$

and for the extension-in-phase of the whole system within any limits

$$\int \ldots \int dp_1 \ldots dq_n \, da_1 \, m_1 \, d\dot{a}_1 \, da_2 \, m_2 \, d\dot{a}_2 \ldots$$

or
$$\int \ldots \int e^{\phi} \, d\epsilon \, da_1 \, m_1 \, d\dot{a}_1 \, da_2 \, m_2 \, d\dot{a}_2 \ldots,$$

or again
$$\int \ldots \int e^{\phi} \, d\mathrm{E} \, da_1 \, m_1 \, d\dot{a}_1 \, da_2 \, m_2 \, d\dot{a}_2 \ldots,$$

since $d\epsilon = d\mathrm{E}$, when a_1, $\dot{a}_1$, a_2, $\dot{a}_2$, etc., are constant. If the limits are expressed by E and $\mathrm{E} + d\mathrm{E}$, a_1 and $a_1 + da_1$, $\dot{a}_1$ and $a_1 + d\dot{a}_1$, etc., the integral reduces to

$$e^{\phi} \, d\mathrm{E} \, da_1 \, m_1 \, d\dot{a}_1 \, da_2 \, m_2 \, d\dot{a}_2 \ldots$$

The values of a_1, $\dot{a}_1$, a_2, $\dot{a}_2$, etc., which make this expression a maximum for constant values of the energy of the whole system and of the differentials $d\mathrm{E}$, da_1, $d\dot{a}_1$, etc., are what may be called the most probable values of a_1, $\dot{a}_1$, etc., in an ensemble in which the whole system is distributed microcanonically. To determine these values we have

$$de^{\phi} = 0,$$

when
$$d(\epsilon + \epsilon_q' + \tfrac{1}{2} m \dot{a}_1^2 + \tfrac{1}{2} m_2 \dot{a}_2^2 + \text{etc.}) = 0.$$

That is,
$$d\phi = 0,$$

when

$$\left(\frac{d\epsilon}{d\phi}\right)_a d\phi + \left(\frac{d\epsilon}{da_1}\right)_{\phi, a} da_1 - A_1' \, da_1 + \text{etc.} + m_1 \dot{a}_1 \, d\dot{a}_1 + \text{etc.} = 0.$$

This requires $\dot{a}_1 = 0, \quad \dot{a}_2 = 0, \quad \text{etc.},$

and $$\left(\frac{d\epsilon}{da_1}\right)_{\phi, a} = A_1', \quad \left(\frac{d\epsilon}{da_2}\right)_{\phi, a} = A_2', \quad \text{etc.}$$

This shows that for any given values of E, a_1, a_2, etc. $\left(\frac{d\epsilon}{da_1}\right)_{\phi, a}$, $\left(\frac{d\epsilon}{da_2}\right)_{\phi, a}$, etc., represent the forces (in the generalized sense) which the external bodies would have to exert to make these values of a_1, a_2, etc., the most probable under the conditions specified. When the differences of the external forces which are exerted by the different systems are negligible, $-(d\epsilon/da_1)_{\phi, a}$, etc., represent these forces.

It is certainly in the quantities relating to a canonical ensemble, $\bar{\epsilon}$, Θ, $\bar{\eta}$, $\bar{A}_1$, etc., a_1, etc., that we find the most complete correspondence with the quantities of the thermodynamic equation (482). Yet the conception itself of the canonical ensemble may seem to some artificial, and hardly germane to a natural exposition of the subject; and the quantities ϵ, $\frac{d\epsilon}{d \log V}$, $\log V$, $\overline{A_1}|_\epsilon$, etc., a_1, etc., or ϵ, $\frac{d\epsilon}{d\phi}$, ϕ, $\left(\frac{d\epsilon}{da_1}\right)_{\phi, a}$, etc., a_1, etc., which are closely related to ensembles of constant energy, and to average and most probable values in such ensembles, and most of which are defined without reference to any ensemble, may appear the most natural analogues of the thermodynamic quantities.

In regard to the naturalness of seeking analogies with the thermodynamic behavior of bodies in canonical or microcanonical ensembles of systems, much will depend upon how we approach the subject, especially upon the question whether we regard energy or temperature as an independent variable.

It is very natural to take energy for an independent variable rather than temperature, because ordinary mechanics furnishes us with a perfectly defined conception of energy, whereas the idea of something relating to a mechanical system and corresponding to temperature is a notion but vaguely defined. Now if the state of a system is given by its energy and the external coördinates, it is incompletely defined, although its partial definition is perfectly clear as far as it goes. The ensemble of phases microcanonically distributed, with the given values of the energy and the external coördinates, will represent the imperfectly defined system better than any other ensemble or single phase. When we approach the subject from this side, our theorems will naturally relate to average values, or most probable values, in such ensembles.

In this case, the choice between the variables of (485) or of (489) will be determined partly by the relative importance which is attached to average and probable values. It would seem that in general average values are the most important, and that they lend themselves better to analytical transformations. This consideration would give the preference to the system of variables in which $\log V$ is the analogue of entropy. Moreover, if we make ϕ the analogue of entropy, we are embarrassed by the necessity of making numerous exceptions for systems of one or two degrees of freedom.

On the other hand, the definition of ϕ may be regarded as a little more simple than that of $\log V$, and if our choice is determined by the simplicity of the definitions of the analogues of entropy and temperature, it would seem that the ϕ system should have the preference. In our definition of these quantities, V was defined first, and e^{ϕ} derived from V by differentiation. This gives the relation of the quantities in the most simple analytical form. Yet so far as the notions are concerned, it is perhaps more natural to regard V as derived from e^{ϕ} by integration. At all events, e^{ϕ} may be defined independently of V, and its definition may be regarded as more simple as not requiring the determination of the zero from which V is measured, which sometimes involves questions of a delicate nature. In fact, the quantity e^{ϕ} may exist, when the definition of V becomes illusory for practical purposes, as the integral by which it is determined becomes infinite.

The case is entirely different, when we regard the temperature as an independent variable, and we have to consider a system which is described as having a certain temperature and certain values for the external coördinates. Here also the state of the system is not completely defined, and will be better represented by an ensemble of phases than by any single phase. What is the nature of such an ensemble as will best represent the imperfectly defined state?

When we wish to give a body a certain temperature, we place it in a bath of the proper temperature, and when we regard what we call thermal equilibrium as established, we say that the body has the same temperature as the bath. Perhaps we place a second body of standard character, which we call a thermometer, in the bath, and say that the first body, the bath, and the thermometer, have all the same temperature.

But the body under such circumstances, as well as the bath, and the thermometer, even if they were entirely isolated from external influences (which it is convenient to suppose in a theoretical discussion), would be continually changing in phase, and in energy as well as in other respects, although our means of observation are not fine enough to perceive these variations.

The series of phases through which the whole system runs in the course of time may not be entirely determined by the energy, but may depend on the initial phase in other respects. In such cases the ensemble obtained by the microcanonical distribution of the whole system, which includes all possible time-ensembles combined in the proportion which seems least arbitrary, will represent better than any one time-ensemble the effect of the bath. Indeed a single time-ensemble, when it is not also a microcanonical ensemble, is too ill-defined a notion to serve the purposes of a general discussion. We will therefore direct our attention, when we suppose the body placed in a bath, to the microcanonical ensemble of phases thus obtained.

If we now suppose the quantity of the substance forming the bath to be increased, the anomalies of the separate energies of the body and of the thermometer in the microcanonical ensemble will be increased, but not without limit. The anomalies of the energy of the bath, considered in comparison with its whole energy, diminish indefinitely as the quantity of the bath is increased, and become in a sense negligible, when the quantity of the bath is sufficiently increased. The ensemble of phases of the body, and of the thermometer, approach a standard form as the quantity of the bath is indefinitely increased. This limiting form is easily shown to be what we have described as the canonical distribution.

Let us write ϵ for the energy of the whole system consisting of the body first mentioned, the bath, and the thermometer (if any), and let us first suppose this system to be distributed canonically with the modulus Θ. We have by (205)

$$\overline{(\epsilon - \bar{\epsilon})^2} = \Theta^2 \frac{d\bar{\epsilon}}{d\Theta},$$

and since

$$\bar{\epsilon}_p = \frac{n}{2}\Theta,$$

$$\frac{d\bar{\epsilon}}{d\Theta} = \frac{n}{2}\frac{d\bar{\epsilon}}{d\bar{\epsilon}_p}.$$

If we write $\Delta\epsilon$ for the anomaly of mean square, we have

$$(\Delta\epsilon)^2 = \overline{(\epsilon - \bar{\epsilon})^2}.$$

If we set

$$\Delta\Theta = \frac{d\Theta}{d\bar{\epsilon}}\Delta\epsilon,$$

$\Delta\Theta$ will represent approximately the increase of Θ which would produce an increase in the average value of the energy equal to its anomaly of mean square. Now these equations give

$$(\Delta\Theta)^2 = \frac{2\Theta^2}{n}\frac{d\bar{\epsilon}_p}{d\bar{\epsilon}},$$

which shows that we may diminish $\Delta\Theta$ indefinitely by increasing the quantity of the bath.

Now our canonical ensemble consists of an infinity of microcanonical ensembles, which differ only in consequence of the different values of the energy which is constant in each. If we consider separately the phases of the first body which occur in the canonical ensemble of the whole system, these phases will form a canonical ensemble of the same modulus. This canonical ensemble of phases of the first body will consist of parts which belong to the different microcanonical ensembles into which the canonical ensemble of the whole system is divided.

Let us now imagine that the modulus of the principal canonical ensemble is increased by $2\Delta\Theta$, and its average energy by $2\Delta\epsilon$. The modulus of the canonical ensemble of the phases of the first body considered separately will be increased by $2\Delta\Theta$. We may regard the infinity of microcanonical ensembles into which we have divided the principal canonical ensemble as each having its energy increased by $2\Delta\epsilon$. Let us see how the ensembles of phases of the first body contained in these microcanonical ensembles are affected. We may assume that they will all be affected in about the same way, as all the differences which come into account may be treated as small. Therefore, the canonical ensemble formed by taking them together will also be affected in the same way. But we know how this is affected. It is by the increase of its modulus by $2\Delta\Theta$, a quantity which vanishes when the quantity of the bath is indefinitely increased.

In the case of an infinite bath, therefore, the increase of the energy of one of the microcanonical ensembles by $2\Delta\epsilon$, produces a vanishing effect on the distribution in energy of the phases of the first body which it contains. But $2\Delta\epsilon$ is more than the average difference of energy between the microcanonical ensembles. The distribution in energy of these phases is therefore the same in the different microcanonical ensembles, and must therefore be canonical, like that of the ensemble which they form when taken together.*

* In order to appreciate the above reasoning, it should be understood that the differences of energy which occur in the canonical ensemble of phases of the first body are not here regarded as vanishing quantities. To fix one's ideas, one may imagine that he has the fineness of perception to make these differences seem large. The difference between the part of these phases which belong to one microcanonical ensemble of the whole system and the part which belongs to another would still be imperceptible, when the quantity of the bath is sufficiently increased.

As a general theorem, the conclusion may be expressed in the words: — If a system of a great number of degrees of freedom is microcanonically distributed in phase, any very small part of it may be regarded as canonically distributed.*

It would seem, therefore, that a canonical ensemble of phases is what best represents, with the precision necessary for exact mathematical reasoning, the notion of a body with a given temperature, if we conceive of the temperature as the state produced by such processes as we actually use in physics to produce a given temperature. Since the anomalies of the body increase with the quantity of the bath, we can only get rid of all that is arbitrary in the ensemble of phases which is to represent the notion of a body of a given temperature by making the bath infinite, which brings us to the canonical distribution.

A comparison of temperature and entropy with their analogues in statistical mechanics would be incomplete without a consideration of their differences with respect to units and zeros, and the numbers used for their numerical specification. If we apply the notions of statistical mechanics to such bodies as we usually consider in thermodynamics, for which the kinetic energy is of the same order of magnitude as the unit of energy, but the number of degrees of freedom is enormous, the values of Θ, $d\epsilon/d\log V$, and $d\epsilon/d\phi$ will be of the same order of magnitude as $1/n$, and the variable part of $\bar{\eta}$, $\log V$, and ϕ will be of the same order of magnitude as n.† If these quantities, therefore, represent in any sense the notions of temperature and entropy, they will nevertheless not be measured in units of the usual order of magnitude, — a fact which must be borne in mind in determining what magnitudes may be regarded as insensible to human observation.

Now nothing prevents our supposing energy and time in our statistical formulae to be measured in such units as may

* It is assumed — and without this assumption the theorem would have no distinct meaning — that the part of the ensemble considered may be regarded as having separate energy.

† See equations (124), (288), (289), and (314); also page 106.

be convenient for physical purposes. But when these units have been chosen, the numerical values of Θ, $d\epsilon/d\log V$, $d\epsilon/d\phi$, $\bar{\eta}$, $\log V$, ϕ, are entirely determined,* and in order to compare them with temperature and entropy, the numerical values of which depend upon an arbitrary unit, we must multiply all values of Θ, $d\epsilon/d\log V$, $d\epsilon/d\phi$ by a constant (K), and divide all values of $\bar{\eta}$, $\log V$, and ϕ by the same constant. This constant is the same for all bodies, and depends only on the units of temperature and energy which we employ. For ordinary units it is of the same order of magnitude as the numbers of atoms in ordinary bodies.

We are not able to determine the numerical value of K, as it depends on the number of molecules in the bodies with which we experiment. To fix our ideas, however, we may seek an expression for this value, based upon very probable assumptions, which will show how we would naturally proceed to its evaluation, if our powers of observation were fine enough to take cognizance of individual molecules.

If the unit of mass of a monatomic gas contains ν atoms, and it may be treated as a system of 3ν degrees of freedom, which seems to be the case, we have for canonical distribution

$$\bar{\epsilon}_p = \tfrac{3}{2}\nu\Theta,$$

$$\frac{d\bar{\epsilon}_p}{d\Theta} = \tfrac{3}{2}\nu. \tag{491}$$

If we write T for temperature, and c_v for the specific heat of the gas for constant volume (or rather the limit toward which this specific heat tends, as rarefaction is indefinitely increased), we have

$$\frac{d\epsilon_p}{dT} = c_v, \tag{492}$$

since we may regard the energy as entirely kinetic. We may set the ϵ_p of this equation equal to the $\bar{\epsilon}_p$ of the preceding,

* The unit of time only affects the last three quantities, and these only by an additive constant, which disappears (with the additive constant of entropy), when differences of entropy are compared with their statistical analogues. See page 19.

where indeed the individual values of which the average is taken would appear to human observation as identical. This gives

$$\frac{d\Theta}{dT} = \frac{2 c_v}{3 \nu},$$

whence
$$\frac{1}{K} = \frac{2 c_v}{3 \nu}. \tag{493}$$

a value recognized by physicists as a constant independent of the kind of monatomic gas considered.

We may also express the value of K in a somewhat different form, which corresponds to the indirect method by which physicists are accustomed to determine the quantity c_v. The kinetic energy due to the motions of the centers of mass of the molecules of a mass of gas sufficiently expanded is easily shown to be equal to

$$\tfrac{3}{2} p v,$$

where p and v denote the pressure and volume. The average value of the same energy in a canonical ensemble of such a mass of gas is

$$\tfrac{3}{2} \Theta \nu,$$

where ν denotes the number of molecules in the gas. Equating these values, we have

$$p v = \Theta \nu, \tag{494}$$

whence
$$\frac{1}{K} = \frac{\Theta}{T} = \frac{p v}{\nu T}. \tag{495}$$

Now the laws of Boyle, Charles, and Avogadro may be expressed by the equation

$$p v = A \nu T, \tag{496}$$

where A is a constant depending only on the units in which energy and temperature are measured. $1/K$, therefore, might be called the constant of the law of Boyle, Charles, and Avogadro as expressed with reference to the true number of molecules in a gaseous body.

Since such numbers are unknown to us, it is more convenient to express the law with reference to relative values. If we denote by M the so-called molecular weight of a gas, that is, a number taken from a table of numbers proportional to the weights of various molecules and atoms, but having one of the values, perhaps the atomic weight of hydrogen, arbitrarily made unity, the law of Boyle, Charles, and Avogadro may be written in the more practical form

$$pv = A' T \frac{m}{M}, \tag{497}$$

where A' is a constant and m the weight of gas considered. It is evident that $1 \quad K$ is equal to the product of the constant of the law in this form and the (true) weight of an atom of hydrogen, or such other atom or molecule as may be given the value unity in the table of molecular weights.

In the following chapter we shall consider the necessary modifications in the theory of equilibrium, when the quantity of matter contained in a system is to be regarded as variable, or, if the system contains more than one kind of matter, when the quantities of the several kinds of matter in the system are to be regarded as independently variable. This will give us yet another set of variables in the statistical equation, corresponding to those of the amplified form of the thermodynamic equation.

CHAPTER XV.

SYSTEMS COMPOSED OF MOLECULES.

THE nature of material bodies is such that especial interest attaches to the dynamics of systems composed of a great number of entirely similar particles, or, it may be, of a great number of particles of several kinds, all of each kind being entirely similar to each other. We shall therefore proceed to consider systems composed of such particles, whether in great numbers or otherwise, and especially to consider the statistical equilibrium of ensembles of such systems. One of the variations to be considered in regard to such systems is a variation in the numbers of the particles of the various kinds which it contains, and the question of statistical equilibrium between two ensembles of such systems relates in part to the tendencies of the various kinds of particles to pass from the one to the other.

First of all, we must define precisely what is meant by statistical equilibrium of such an ensemble of systems. The essence of statistical equilibrium is the permanence of the number of systems which fall within any given limits with respect to phase. We have therefore to define how the term "phase" is to be understood in such cases. If two phases differ only in that certain entirely similar particles have changed places with one another, are they to be regarded as identical or different phases? If the particles are regarded as indistinguishable, it seems in accordance with the spirit of the statistical method to regard the phases as identical. In fact, it might be urged that in such an ensemble of systems as we are considering no identity is possible between the particles of different systems except that of qualities, and if ν particles of one system are described as entirely similar to one another and to ν of another system, nothing remains on which to base the indentification of any particular particle of the first system with any particular particle of the second. And this would be true, if the ensemble of systems had a simultaneous objective existence. But it hardly applies to the creations of the imagination. In the cases which we have been considering, and in those which we shall consider, it is not only possible to conceive of the motion of an ensemble of similar systems simply as possible cases of the motion of a single system, but it is actually in large measure for the sake of representing more clearly the possible cases of the motion of a single system that we use the conception of an ensemble of systems. The perfect similarity of several particles of a system will not in the least interfere with the identification of a particular particle in one case with a particular particle in another. The question is one to be decided in accordance with the requirements of practical convenience in the discussion of the problems with which we are engaged.

Our present purpose will often require us to use the terms *phase*, *density-in-phase*, *statistical equilibrium*, and other connected terms on the supposition that phases are *not* altered by the exchange of places between similar particles. Some of the most important questions with which we are concerned have reference to phases thus defined. We shall call them phases determined by generic definitions, or briefly, generic phases. But we shall also be obliged to discuss phases defined by the narrower definition (so that exchange of position between similar particles is regarded as changing the phase), which will be called phases determined by specific definitions, or briefly, specific phases. For the analytical description of a specific phase is more simple than that of a generic phase. And it is a more simple matter to make a multiple integral extend over all possible specific phases than to make one extend without repetition over all possible generic phases.

It is evident that if $\nu_1, \nu_2 \ldots \nu_h$, are the numbers of the different kinds of molecules in any system, the number of specific phases embraced in one generic phase is represented by the continued product $\underline{|\nu_1}\, \underline{|\nu_2} \ldots \underline{|\nu_h}$, and the coefficient of probability of a generic phase is the sum of the probability-coefficients of the specific phases which it represents. When these are equal among themselves, the probability-coefficient of the generic phase is equal to that of the specific phase multiplied by $|\underline{\nu_1}\,|\underline{\nu_2} \ldots |\underline{\nu_h}$. It is also evident that statistical equilibrium may subsist with respect to generic phases without statistical equilibrium with respect to specific phases, but not *vice versa.*

Similar questions arise where one particle is capable of several equivalent positions. Does the change from one of these positions to another change the phase? It would be most natural and logical to make it affect the specific phase, but not the generic. The number of specific phases contained in a generic phase would then be $|\underline{\nu_1}\,\kappa_1^{\nu_1} \ldots |\underline{\nu_h}\,\kappa_h^{\nu_h}$, where $\kappa_1, \ldots \kappa_h$ denote the numbers of equivalent positions belonging to the several kinds of particles. The case in which a κ is infinite would then require especial attention. It does not appear that the resulting complications in the formulae would be compensated by any real advantage. The reason of this is that in problems of real interest equivalent positions of a particle will always be equally probable. In this respect, equivalent positions of the same particle are entirely unlike the $|\underline{\nu}$ different ways in which ν particles may be distributed in ν different positions. Let it therefore be understood that in spite of the physical equivalence of different positions of the same particle they are to be considered as constituting a difference of generic phase as well as of specific. The number of specific phases contained in a generic phase is therefore always given by the product $|\underline{\nu_1}\,|\underline{\nu_2} \ldots |\underline{\nu_h}$.

Instead of considering, as in the preceding chapters, ensembles of systems differing only in phase, we shall now suppose that the systems constituting an ensemble are composed of particles of various kinds, and that they differ not only in phase but also in the numbers of these particles which they contain. The external coördinates of all the systems in the ensemble are supposed, as heretofore, to have the same value, and when they vary, to vary together. For distinction, we may call such an ensemble a *grand ensemble*, and one in which the systems differ only in phase a *petit ensemble.* A grand ensemble is therefore composed of a multitude of petit ensembles. The ensembles which we have hitherto discussed are petit ensembles.

Let $\nu_1, \ldots \nu_h$, etc., denote the numbers of the different kinds of particles in a system, ϵ its energy, and $q_1, \ldots q_n$, $p_1, \ldots p_n$ its coördinates and momenta. If the particles are of the nature of material points, the number of coördinates (n) of the system will be equal to $3\nu_1 \ldots + 3\nu_h$. But if the particles are less simple in their nature, if they are to be treated as rigid solids, the orientation of which must be regarded, or if they consist each of several atoms, so as to have more than three degrees of freedom, the number of coördinates of the system will be equal to the sum of ν_1, ν_2, etc., multiplied each by the number of degrees of freedom of the kind of particle to which it relates.

Let us consider an ensemble in which the number of systems having $\nu_1, \ldots \nu_h$ particles of the several kinds, and having values of their coördinates and momenta lying between the limits q_1 and $q_1 + dq_1$, p_1 and $p_1 + dp_1$, etc., is represented by the expression

$$\frac{N e^{\frac{\Omega + \mu_1 \nu_1 \ldots + \mu_h \nu_h - \epsilon}{\Theta}}}{\lfloor \nu_1 \ldots \lfloor \nu_h} dp_1 \ldots dq_n, \qquad (498)$$

where $N, \Omega, \Theta, \mu_1, \ldots \mu_h$ are constants, N denoting the total number of systems in the ensemble. The expression

$$\frac{N e^{\frac{\Omega + \mu_1 \nu_1 \ldots + \mu_h \nu_h - \epsilon}{\Theta}}}{\lfloor \nu_1 \ldots \lfloor \nu_h} \qquad (499)$$

evidently represents the density-in-phase of the ensemble within the limits described, that is, for a phase specifically defined. The expression

$$\frac{e^{\frac{\Omega + \mu_1 \nu_1 \ldots + \mu_h \nu_h - \epsilon}{\Theta}}}{\lfloor \nu_1 \ldots \lfloor \nu_h} \qquad (500)$$

is therefore the probability-coefficient for a phase specifically defined. This has evidently the same value for all the $\lfloor\nu_1 \ldots \lfloor\nu_h$ phases obtained by interchanging the phases of particles of the same kind. The probability-coefficient for a generic phase will be $\lfloor\nu_1 \ldots \lfloor\nu_h$ times as great, viz.,

$$e^{\frac{\Omega+\mu_1\nu_1\ldots+\mu_h\nu_h-\epsilon}{\Theta}}. \tag{501}$$

We shall say that such an ensemble as has been described is *canonically distributed*, and shall call the constant Θ its modulus. It is evidently what we have called a grand ensemble. The petit ensembles of which it is composed are canonically distributed, according to the definitions of Chapter IV, since the expression

$$\frac{e^{\frac{\Omega+\mu_1\nu_1\ldots+\mu_h\nu_h}{\Theta}}}{\lfloor\nu_1 \ldots \lfloor\nu_h} \tag{502}$$

is constant for each petit ensemble. The grand ensemble, therefore, is in statistical equilibrium with respect to specific phases.

If an ensemble, whether grand or petit, is identical so far as generic phases are concerned with one canonically distributed, we shall say that its distribution is canonical with respect to generic phases. Such an ensemble is evidently in statistical equilibrium with respect to generic phases, although it may not be so with respect to specific phases.

If we write H for the index of probability of a generic phase in a grand ensemble, we have for the case of canonical distribution

$$\mathrm{H} = \frac{\Omega + \mu_1\nu_1 \ldots + \mu_h\nu_h - \epsilon}{\Theta}. \tag{503}$$

It will be observed that the H is a linear function of ϵ and $\nu_1, \ldots \nu_h$; also that whenever the index of probability of generic phases in a grand ensemble is a linear function of $\epsilon, \nu_1, \ldots \nu_h$, the ensemble is canonically distributed with respect to generic phases.

The constant Ω we may regard as determined by the equation

$$N = \Sigma_{\nu_1} \ldots \Sigma_{\nu_h} \int \underset{\text{phases}}{\overset{\text{all}}{\ldots}} \int \frac{N e^{\frac{\Omega + \mu_1 \nu_1 \ldots \mu_h \nu_h - \epsilon}{\Theta}}}{\lfloor \nu_1 \ldots \lfloor \nu_h} \, dp_1 \ldots dq_n, \tag{504}$$

or

$$e^{-\frac{\Omega}{\Theta}} = \Sigma_{\nu_1} \ldots \Sigma_{\nu_h} \frac{e^{\frac{\mu_1 \nu_1 \ldots \mu_h \nu_h}{\Theta}}}{\lfloor \nu_1 \ldots \lfloor \nu_h} \int \underset{\text{phases}}{\overset{\text{all}}{\ldots}} \int e^{-\frac{\epsilon}{\Theta}} \, dp_1 \ldots dq_n, \tag{505}$$

where the multiple sum indicated by $\Sigma_{\nu_1} \ldots \Sigma_{\nu_h}$ includes all terms obtained by giving to each of the symbols $\nu_1 \ldots \nu_h$ all integral values from zero upward, and the multiple integral (which is to be evaluated separately for each term of the multiple sum) is to be extended over all the (specific) phases of the system having the specified numbers of particles of the various kinds. The multiple integral in the last equation is what we have represented by $e^{-\frac{\psi}{\Theta}}$. See equation (92). We may therefore write

$$e^{-\frac{\Omega}{\Theta}} = \Sigma_{\nu_1} \ldots \Sigma_{\nu_h} \frac{e^{\frac{\mu_1 \nu_1 \ldots \mu_h \nu_h - \psi}{\Theta}}}{\lfloor \nu_1 \ldots \lfloor \nu_h}. \tag{506}$$

It should be observed that the summation includes a term in which all the symbols $\nu_1 \ldots \nu_h$ have the value zero. We must therefore recognize in a certain sense a system consisting of no particles, which, although a barren subject of study in itself, cannot well be excluded as a particular case of a system of a variable number of particles. In this case ϵ is constant, and there are no integrations to be performed. We have therefore*

$$e^{-\frac{\psi}{\Theta}} = e^{-\frac{\epsilon}{\Theta}}, \quad i.e., \quad \psi = \epsilon.$$

* This conclusion may appear a little strained. The original definition of ψ may not be regarded as fairly applying to systems of no degrees of freedom. We may therefore prefer to regard these equations as defining ψ in this case.

The value of ϵ_p is of course zero in this case. But the value of ϵ_q contains an arbitrary constant, which is generally determined by considerations of convenience, so that ϵ_q and ϵ do not necessarily vanish with $\nu_1, \dots \nu_h$.

Unless $-\Omega$ has a finite value, our formulae become illusory. We have already, in considering petit ensembles canonically distributed, found it necessary to exclude cases in which $-\psi$ has not a finite value.* The same exclusion would here make $-\psi$ finite for any finite values of $\nu_1 \dots \nu_h$. This does not necessarily make a multiple series of the form (506) finite. We may observe, however, that if for all values of $\nu_1 \dots \nu_h$

$$-\psi \leqq c_0 + c_1 \nu_1, \dots + c_h \nu_h, \tag{507}$$

where $c_0, c_1, \dots c_h$ are constants or functions of Θ,

$$e^{-\frac{\Omega}{\Theta}} \leqq \Sigma_{\nu_1} \dots \Sigma_{\nu_h} \frac{e^{\frac{c_0+(\mu_1+c_1)\nu_1 \dots +(\mu_h+c_h)\nu_h}{\Theta}}}{\underline{|\nu_1} \dots \underline{|\nu_h}}$$

i. e.,
$$e^{-\frac{\Omega}{\Theta}} \leqq e^{\frac{c_0}{\Theta}} \Sigma_{\nu_1} \frac{e^{\frac{\mu_1+c_1}{\Theta}\nu_1}}{\underline{|\nu_1}} \dots \Sigma_{\nu_h} \frac{e^{\frac{\mu_h+c_h}{\Theta}\nu_h}}{\underline{|\nu_h}}$$

i. e.,
$$e^{-\frac{\Omega}{\Theta}} \leqq e^{\frac{c_0}{\Theta}} e^{e^{\frac{\mu_1+c_1}{\Theta}}} \dots e^{e^{\frac{\mu_h+\nu_h}{\Theta}}}$$

i. e.,
$$-\frac{\Omega}{\Theta} \leqq \frac{c_0}{\Theta} + e^{\frac{\mu_1+c_1}{\Theta}} \dots + e^{\frac{\mu_h+c_h}{\Theta}}. \tag{508}$$

The value of $-\Omega$ will therefore be finite, when the condition (507) is satisfied. If therefore we assume that $-\Omega$ is finite, we do not appear to exclude any cases which are analogous to those of nature.†

The interest of the ensemble which has been described lies in the fact that it may be in statistical equilbrium, both in

* See Chapter IV, page 35.

† If the external coördinates determine a certain volume within which the system is confined, the contrary of (507) would imply that we could obtain an infinite amount of work by crowding an infinite quantity of matter into a finite volume.

respect to exchange of energy and exchange of particles, with other grand ensembles canonically distributed and having the same values of Θ and of the coefficients μ_1, μ_2, etc., when the circumstances are such that exchange of energy and of particles are possible, and when equilibrium would not subsist, were it not for equal values of these constants in the two ensembles.

With respect to the exchange of energy, the case is exactly the same as that of the petit ensembles considered in Chapter IV, and needs no especial discussion. The question of exchange of particles is to a certain extent analogous, and may be treated in a somewhat similar manner. Let us suppose that we have two grand ensembles canonically distributed with respect to specific phases, with the same value of the modulus and of the coefficients $\mu_1 \ldots \mu_h$, and let us consider the ensemble of all the systems obtained by combining each system of the first ensemble with each of the second.

The probability-coefficient of a generic phase in the first ensemble may be expressed by

$$e^{\frac{\Omega' + \mu_1 \nu_1' \ldots + \mu_h \nu_h' - \epsilon'}{\Theta}} \tag{509}$$

The probability-coefficient of a specific phase will then be expressed by

$$\frac{e^{\frac{\Omega' + \mu_1 \nu_1' \ldots + \mu_h \nu_h' - \epsilon'}{\Theta}}}{\lfloor \nu_1' \ldots \lfloor \nu_h'}, \tag{510}$$

since each generic phase comprises $\lfloor \nu_1 \ldots \lfloor \nu_h$ specific phases. In the second ensemble the probability-coefficients of the generic and specific phases will be

$$e^{\frac{\Omega'' + \mu_1 \nu_1'' \ldots + \mu_h \nu_h'' - \epsilon''}{\Theta}}, \tag{511}$$

and

$$\frac{e^{\frac{\Omega'' + \mu_1 \nu_1'' \ldots + \mu_h \nu_h'' - \epsilon''}{\Theta}}}{\lfloor \nu_1'' \ldots \lfloor \nu_h''}. \tag{512}$$

The probability-coefficient of a generic phase in the third ensemble, which consists of systems obtained by regarding each system of the first ensemble combined with each of the second as forming a system, will be the product of the probability-coefficients of the generic phases of the systems combined, and will therefore be represented by the formula

$$e^{\frac{\Omega'''+\mu_1\nu_1'''\ldots+\mu_h\nu_h'''-\epsilon'''}{\Theta}} \tag{513}$$

where $\Omega''' = \Omega' + \Omega''$, $\epsilon''' = \epsilon' + \epsilon''$, $\nu_1''' = \nu_1' + \nu_1''$, etc. It will be observed that ν_1''', etc., represent the numbers of particles of the various kinds in the third ensemble, and ϵ''' its energy; also that Ω''' is a constant. The third ensemble is therefore canonically distributed with respect to generic phases.

If all the systems in the same generic phase in the third ensemble were equably distributed among the $\lfloor\nu_1''' \ldots \lfloor\nu_h'''$ specific phases which are comprised in the generic phase, the probability-coefficient of a specific phase would be

$$\frac{e^{\frac{\Omega'''+\mu_1\nu_1'''\ldots+\mu_h\nu_h'''-\epsilon'''}{\Theta}}}{\lfloor\nu_1''' \ldots \lfloor\nu_h'''}. \tag{514}$$

In fact, however, the probability-coefficient of any specific phase which occurs in the third ensemble is

$$\frac{e^{\frac{\Omega'''+\mu_1\nu_1'''\ldots+\mu_h\nu_h'''-\epsilon'''}{\Theta}}}{\lfloor\nu_1' \ldots \lfloor\nu_h' \lfloor\nu_1'' \ldots \lfloor\nu_h''}, \tag{515}$$

which we get by multiplying the probability-coefficients of specific phases in the first and second ensembles. The difference between the formulae (514) and (515) is due to the fact that the generic phases to which (513) relates include not only the specific phases occurring in the third ensemble and having the probability-coefficient (515), but also all the specific phases obtained from these by interchange of similar particles between two combined systems. Of these the probability-coefficient is evidently zero, as they do not occur in the ensemble.

Now this third ensemble is in statistical equilibrium, with respect both to specific and generic phases, since the ensembles from which it is formed are so. This statistical equilibrium is not dependent on the equality of the modulus and the co-efficients $\mu_1, \ldots \mu_h$ in the first and second ensembles. It depends only on the fact that the two original ensembles were separately in statistical equilibrium, and that there is no interaction between them, the combining of the two ensembles to form a third being purely nominal, and involving no physical connection. This independence of the systems, determined physically by forces which prevent particles from passing from one system to the other, or coming within range of each other's action, is represented mathematically by infinite values of the energy for particles in a space dividing the systems. Such a space may be called a diaphragm.

If we now suppose that, when we combine the systems of the two original ensembles, the forces are so modified that the energy is no longer infinite for particles in all the space forming the diaphragm, but is diminished in a part of this space, so that it is possible for particles to pass from one system to the other, this will involve a change in the function ϵ''' which represents the energy of the combined systems, and the equation $\epsilon''' = \epsilon' + \epsilon''$ will no longer hold. Now if the coefficient of probability in the third ensemble were represented by (513) with this new function ϵ''', we should have statistical equilibrium, with respect to generic phases, although not to specific. But this need involve only a trifling change in the distribution of the third ensemble,* a change represented by the addition of comparatively few systems in which the transference of particles is taking place to the immense number

* It will be observed that, so far as the distribution is concerned, very large and infinite values of ϵ (for certain phases) amount to nearly the same thing, — one representing the total and the other the nearly total exclusion of the phases in question. An infinite change, therefore, in the value of ϵ (for certain phases) may represent a vanishing change in the distribution.

obtained by combining the two original ensembles. The difference between the ensemble which would be in statistical equilibrium, and that obtained by combining the two original ensembles may be diminished without limit, while it is still possible for particles to pass from one system to another. In this sense we may say that the ensemble formed by combining the two given ensembles may still be regarded as in a state of (approximate) statistical equilibrium with respect to generic phases, when it has been made possible for particles to pass between the systems combined, and when statistical equilibrium for specific phases has therefore entirely ceased to exist, and when the equilibrium for generic phases would also have entirely ceased to exist, if the given ensembles had not been canonically distributed, with respect to generic phases, with the same values of Θ and $\mu_1, \ldots \mu_h$.

It is evident also that considerations of this kind will apply separately to the several kinds of particles. We may diminish the energy in the space forming the diaphragm for one kind of particle and not for another. This is the mathematical expression for a "semipermeable" diaphragm. The condition necessary for statistical equilibrium where the diaphragm is permeable only to particles to which the suffix $(\ )_1$ relates will be fulfilled when μ_1 and Θ have the same values in the two ensembles, although the other coefficients μ_2, etc., may be different.

This important property of grand ensembles with canonical distribution will supply the motive for a more particular examination of the nature of such ensembles, and especially of the comparative numbers of systems in the several petit ensembles which make up a grand ensemble, and of the average values in the grand ensemble of some of the most important quantities, and of the average squares of the deviations from these average values.

The probability that a system taken at random from a grand ensemble canonically distributed will have exactly $\nu_1, \ldots \nu_h$ particles of the various kinds is expressed by the multiple integral

$$\int_{\text{phases}}^{\text{all}} \cdots \int \frac{e^{\frac{\Omega+\mu_1\nu_1\ldots+\mu_h\nu_h-\epsilon}{\Theta}}}{\lfloor\nu_1 \ldots \lfloor\nu_h} dp_1 \ldots dq_n \tag{516}$$

or

$$\frac{e^{\frac{\Omega+\mu_1\nu_1\ldots+\mu_h\nu_h-\psi}{\Theta}}}{\lfloor\nu_1 \ldots \lfloor\nu_h}. \tag{517}$$

This may be called the probability of the petit ensemble $(\nu_1, \ldots \nu_h)$. The sum of all such probabilities is evidently unity. That is,

$$\Sigma_{\nu_1} \ldots \Sigma_{\nu_h} \frac{e^{\frac{\Omega+\mu_1\nu_1\ldots+\mu_h\nu_h-\psi}{\Theta}}}{\lfloor\nu_1 \ldots \lfloor\nu_h} = 1, \tag{518}$$

which agrees with (506).

The average value in the grand ensemble of any quantity u, is given by the formula

$$\overline{u} = \Sigma_{\nu_1} \ldots \Sigma_{\nu_h} \int_{\text{phases}}^{\text{all}} \cdots \int \frac{u\, e^{\frac{\Omega+\mu_1\nu_1\ldots+\mu_h\nu_h-\epsilon}{\Theta}}}{\lfloor\nu_1 \ldots \lfloor\nu_h} dp_1 \ldots dq_n. \tag{519}$$

If u is a function of $\nu_1, \ldots \nu_h$ alone, *i. e.*, if it has the same value in all systems of any same petit ensemble, the formula reduces to

$$\overline{u} = \Sigma_{\nu_1} \ldots \Sigma_{\nu_h} \frac{u\, e^{\frac{\Omega+\mu_1\nu_1\ldots+\mu_h\nu_h-\psi}{\Theta}}}{\lfloor\nu_1 \ldots \lfloor\nu_h}. \tag{520}$$

Again, if we write $\overline{u}\rfloor_{\text{grand}}$ and $\overline{u}\rfloor_{\text{petit}}$ to distinguish averages in the grand and petit ensembles, we shall have

$$\overline{u}\rfloor_{\text{grand}} = \Sigma_{\nu_1} \ldots \Sigma_{\nu_h} \overline{u}\rfloor_{\text{petit}} \frac{e^{\frac{\Omega+\mu_1\nu_1\ldots+\mu_h\nu_h-\psi}{\Theta}}}{\lfloor\nu_1 \ldots \lfloor\nu_h}. \tag{521}$$

In this chapter, in which we are treating of grand ensembles, $\overline{u}$ will always denote the average for a grand ensemble. In the preceding chapters, $\overline{u}$ has always denoted the average for a petit ensemble.

Equation (505), which we repeat in a slightly different form, viz.,

$$e^{-\frac{\Omega}{\Theta}} = \Sigma_{\nu_1} \ldots \Sigma_{\nu_h} \int \ldots \int_{\text{phases}}^{\text{all}} \frac{e^{\frac{\mu_1\nu_1 \ldots + \mu_h\nu_h - \epsilon}{\Theta}}}{\lfloor \nu_1 \ldots \lfloor \nu_h} dp_1 \ldots dq_n, \quad (522)$$

shows that Ω is a function of Θ and $\mu_1, \ldots \mu_h$; also of the external coördinates a_1, a_2, etc., which are involved implicitly in ϵ. If we differentiate the equation regarding all these quantities as variable, we have

$$e^{-\frac{\Omega}{\Theta}}\left(-\frac{d\Omega}{\Theta} + \frac{\Omega}{\Theta^2}d\Theta\right) =$$

$$-\frac{d\Theta}{\Theta^2}\Sigma_{\nu_1}\ldots\Sigma_{\nu_h}\int\ldots\int_{\text{phases}}^{\text{all}} \frac{(\mu_1\nu_1\ldots + \mu_h\nu_h - \epsilon)\, e^{\frac{\mu_1\nu_1\ldots+\mu_h\nu_h-\epsilon}{\Theta}}}{\lfloor \nu_1 \ldots \lfloor \nu_h} dp_1\ldots dq_n$$

$$+\frac{d\mu_1}{\Theta}\Sigma_{\nu_1}\ldots\Sigma_{\nu_h}\int\ldots\int_{\text{phases}}^{\text{all}} \frac{\nu_1 e^{\frac{\mu_1\nu_1\ldots+\mu_h\nu_h-\epsilon}{\Theta}}}{\lfloor \nu_1 \ldots \lfloor \nu_h} dp_1\ldots dq_n$$

$$+ \text{ etc.}$$

$$-\frac{da_1}{\Theta}\Sigma_{\nu_1}\ldots\Sigma_{\nu_h}\int\ldots\int_{\text{phases}}^{\text{all}} \frac{d\epsilon}{da_1}\frac{e^{\frac{\mu_1\nu_1\ldots+\mu_h\nu_h-\epsilon}{\Theta}}}{\lfloor \nu_1 \ldots \lfloor \nu_h} dp_1\ldots dq_n,$$

$$- \text{ etc.} \quad (523)$$

If we multiply this equation by $e^{\frac{\Omega}{\Theta}}$, and set as usual A_1, A_2, etc., for $-d\epsilon/da_1$, $-d\epsilon/da_2$, etc., we get in virtue of the law expressed by equation (519),

$$-\frac{d\Omega}{\Theta} + \frac{\Omega}{\Theta^2}d\Theta = -\frac{d\Theta}{\Theta^2}(\mu_1\bar{\nu}_1 \ldots + \mu_h\bar{\nu}_h - \bar{\epsilon})$$

$$+\frac{d\mu_1}{\Theta}\bar{\nu}_1 + \frac{d\mu_2}{\Theta}\bar{\nu}_2 + \text{ etc.}$$

$$+\frac{da_1}{\Theta}\bar{A}_1 + \frac{da_2}{\Theta}\bar{A}_2 + \text{ etc.}; \quad (524)$$

that is,

$$d\Omega = \frac{\Omega + \mu_1 \bar{\nu}_1 \ldots \mu_h \bar{\nu}_h - \bar{\epsilon}}{\Theta} d\Theta - \Sigma \bar{\nu}_1 \, d\mu_1 - \Sigma \bar{A}_1 \, da_1. \tag{525}$$

Since equation (503) gives

$$\frac{\Omega + \mu_1 \bar{\nu}_1 \ldots + \mu_h \bar{\nu}_h - \bar{\epsilon}}{\Theta} = \bar{\mathrm{H}}, \tag{526}$$

the preceding equation may be written

$$d\Omega = \bar{\mathrm{H}} \, d\Theta - \Sigma \bar{\nu}_1 \, d\mu_1 - \Sigma \bar{A}_1 \, da_1. \tag{527}$$

Again, equation (526) gives

$$d\Omega + \Sigma \mu_1 \, d\bar{\nu}_1 + \Sigma \bar{\nu}_1 \, d\mu_1 - d\bar{\epsilon} = \Theta \, d\bar{\mathrm{H}} + \bar{\mathrm{H}} \, d\Theta. \tag{528}$$

Eliminating $d\Omega$ from these equations, we get

$$d\bar{\epsilon} = - \Theta \, d\bar{\mathrm{H}} + \Sigma \mu_1 \, d\bar{\nu}_1 - \Sigma \bar{A}_1 \, da_1. \tag{529}$$

If we set

$$\Psi = \bar{\epsilon} + \Theta \bar{\mathrm{H}}, \tag{530}$$

$$d\Psi = d\epsilon + \Theta \, d\bar{\mathrm{H}} + \bar{\mathrm{H}} \, d\Theta, \tag{531}$$

we have

$$d\Psi = \bar{\mathrm{H}} \, d\Theta + \Sigma \mu_1 \, d\bar{\nu}_1 - \Sigma \bar{A}_1 \, da_1. \tag{532}$$

The corresponding thermodynamic equations are

$$d\epsilon = T \, d\eta + \Sigma \mu_1 dm_1 - \Sigma A_1 \, da_1, \tag{533}$$

$$\psi = \epsilon - T \eta, \tag{534}$$

$$d\psi = - \eta \, dT + \Sigma \mu_1 \, dm_1 - \Sigma A_1 \, da_1. \tag{535}$$

These are derived from the thermodynamic equations (114) and (117) by the addition of the terms necessary to take account of variation in the quantities (m_1, m_2, etc.) of the several substances of which a body is composed. The correspondence of the equations is most perfect when the component substances are measured in such units that m_1, m_2, etc., are proportional to the numbers of the different kinds of molecules or atoms. The quantities μ_1, μ_2, etc., in these thermodynamic equations may be defined as differential coefficients by either of the equations in which they occur.*

* Compare Transactions Connecticut Academy, Vol. III, pages 116 ff.

If we compare the statistical equations (529) and (532) with (114) and (112), which are given in Chapter IV, and discussed in Chapter XIV, as analogues of thermodynamic equations, we find considerable difference. Beside the terms corresponding to the additional terms in the thermodynamic equations of this chapter, and beside the fact that the averages are taken in a grand ensemble in one case and in a petit in the other, the analogues of entropy, H and η, are quite different in definition and value. We shall return to this point after we have determined the order of magnitude of the usual anomalies of $\nu_1, \ldots \nu_h$.

If we differentiate equation (518) with respect to μ_1, and multiply by Θ, we get

$$\Sigma_{\nu_1} \ldots \Sigma_{\nu_h} \left(\frac{d\Omega}{d\mu_1} + \nu_1 \right) \frac{e^{\frac{\Omega + \mu_1\nu_1 \ldots + \mu_h\nu_h - \psi}{\Theta}}}{\lfloor \nu_1 \ldots \lfloor \nu_h} = 0, \tag{536}$$

whence $d\Omega/d\mu_1 = -\bar{\nu}_1$, which agrees with (527). Differentiating again with respect to μ_1, and to μ_2, and setting

$$\frac{d\Omega}{d\mu_1} = -\bar{\nu}_1, \quad \frac{d\Omega}{d\mu_2} = -\bar{\nu}_2,$$

we get

$$\Sigma_{\nu_1} \ldots \Sigma_{\nu_h} \left(\frac{d^2\Omega}{d{\mu_1}^2} + \frac{(\nu_1 - \bar{\nu}_1)^2}{\Theta} \right) \frac{e^{\frac{\Omega + \mu_1\nu_1 \ldots + \mu_h\nu_h - \psi}{\Theta}}}{\lfloor \nu_1 \ldots \lfloor \nu_h} = 0, \tag{537}$$

$$\Sigma_{\nu_1} \ldots \Sigma_{\nu_h} \left(\frac{d^2\Omega}{d\mu_1\, d\mu_2} + \frac{(\nu_1 - \bar{\nu}_1)(\nu_2 - \bar{\nu}_2)}{\Theta} \right) \frac{e^{\frac{\Omega + \mu_1\nu_1 \ldots + \mu_h\nu_h - \psi}{\Theta}}}{\lfloor \nu_1 \ldots \lfloor \nu_h} = 0. \tag{538}$$

The first members of these equations represent the average values of the quantities in the principal parentheses. We have therefore

$$\overline{(\nu_1 - \bar{\nu}_1)^2} = \overline{\nu_1^2} - \bar{\nu}_1^2 = -\Theta \frac{d^2\Omega}{d{\mu_1}^2} = \Theta \frac{d\bar{\nu}_1}{d\mu_1}, \tag{539}$$

$$\overline{(\nu_1 - \bar{\nu}_1)(\nu_2 - \bar{\nu}_2)} = \overline{\nu_1 \nu_2} - \bar{\nu}_1 \bar{\nu}_2 = -\Theta \frac{d^2\Omega}{d\mu_1\, d\mu_2} = \Theta \frac{d\bar{\nu}_1}{d\mu_2} = \Theta \frac{d\bar{\nu}_2}{d\mu_1}. \tag{540}$$

From equation (539) we may get an idea of the order of magnitude of the divergences of ν_1 from its average value in the ensemble, when that average value is great. The equation may be written

$$\frac{\overline{(\nu_1 - \overline{\nu_1})^2}}{\overline{\nu_1}^2} = \frac{\Theta}{\overline{\nu_1}^2} \frac{d\overline{\nu_1}}{d\mu_1}, \tag{541}$$

The second member of this equation will in general be small when $\overline{\nu}_1$ is great. Large values are not necessarily excluded, but they must be confined within very small limits with respect to μ. For if

$$\frac{\overline{(\nu_1 - \overline{\nu_1})^2}}{\overline{\nu_1}^2} > \frac{1}{\overline{\nu_1}^{\frac{1}{2}}}, \tag{542}$$

for all values of μ_1 between the limits μ_1' and μ_1'', we shall have between the same limits

$$\frac{\Theta}{\overline{\nu_1}^{\frac{3}{2}}} d\overline{\nu_1} > d\mu_1, \tag{543}$$

and therefore

$$\tfrac{1}{2}\Theta\left(\frac{1}{\overline{\nu_1}'^{\frac{1}{2}}} - \frac{1}{\overline{\nu_1}''^{\frac{1}{2}}}\right) > \mu_1'' - \mu_1'. \tag{544}$$

The difference $\mu_1'' - \mu_1'$ is therefore numerically a very small quantity. To form an idea of the importance of such a difference, we should observe that in formula (498) μ_1 is multiplied by ν_1 and the product subtracted from the energy. A very small difference in the value of μ_1 may therefore be important. But since $\nu\Theta$ is always less than the kinetic energy of the system, our formula shows that $\mu_1'' - \mu_1'$, even when multiplied by $\overline{\nu}_1'$ or $\overline{\nu}_1''$, may still be regarded as an insensible quantity.

We can now perceive the leading characteristics with respect to properties sensible to human faculties of such an ensemble as we are considering (a grand ensemble canonically distributed), when the average numbers of particles of the various kinds are of the same order of magnitude as the number of molecules in the bodies which are the subject of physical experiment. Although the ensemble contains systems having the widest possible variations in respect to the numbers of the particles which they contain, these variations are practically contained within such narrow limits as to be insensible, except for particular values of the constants of the ensemble. This exception corresponds precisely to the case of nature, when certain thermodynamic quantities corresponding to Θ, μ_1, μ_2, etc., which in general determine the separate densities of various components of a body, have certain values which make these densities indeterminate, in other words, when the conditions are such as determine coexistent phases of matter. Except in the case of these particular values, the grand ensemble would not differ to human faculties of perception from a petit ensemble, viz., any one of the petit ensembles which it contains in which $\bar{\nu}_1$, $\bar{\nu}_2$, etc., do not sensibly differ from their average values.

Let us now compare the quantities H and η, the average values of which (in a grand and a petit ensemble respectively) we have seen to correspond to entropy. Since

$$\mathrm{H} = \frac{\Omega + \mu_1 \nu_1 \ldots + \mu_h \nu_h - \epsilon}{\Theta},$$

and

$$\eta = \frac{\psi - \epsilon}{\Theta},$$

$$\mathrm{H} - \eta = \frac{\Omega + \mu_1 \nu_1 \ldots + \mu_h \nu_h - \psi}{\Theta}. \tag{545}$$

A part of this difference is due to the fact that H relates to generic phases and η to specific. If we write η_{gen} for the index of probability for generic phases in a petit ensemble, we have

$$\eta_{\mathrm{gen}} = \eta + \log \underline{|\nu_1} \ldots \underline{|\nu_h}, \tag{546}$$

$$\mathrm{H} - \eta = \mathrm{H} - \eta_{\mathrm{gen}} + \log \underline{|\nu_1} \ldots \underline{|\nu_h}, \tag{547}$$

$$\mathrm{H} - \eta_{\mathrm{gen}} = \frac{\Omega + \mu_1 \nu_1 \ldots + \mu_h \nu_h - \psi}{\Theta} - \log \underline{|\nu_1} \ldots \underline{|\nu_h}. \tag{548}$$

This is the logarithm of the probability of the petit ensemble $(\nu_1 \ldots \nu_h)$.* If we set

* See formula (517).

$$\frac{\psi_{gen} - \epsilon}{\Theta} = \eta_{gen}, \tag{549}$$

which corresponds to the equation

$$\frac{\psi - \epsilon}{\Theta} = \eta, \tag{550}$$

we have $\qquad \psi_{gen} = \psi + \Theta \log \lfloor\nu_1 \ldots \lfloor\nu_h,$

and $\qquad H - \eta_{gen} = \frac{\Omega + \mu_1\nu_1 \ldots + \mu_h\nu_h - \psi_{gen}}{\Theta}. \qquad (551)$

This will have a maximum when *

$$\frac{d\psi_{gen}}{d\nu_1} = \mu_1, \qquad \frac{d\psi_{gen}}{d\nu_2} = \mu_2, \qquad \text{etc.} \tag{552}$$

Distinguishing values corresponding to this maximum by accents, we have approximately, when $\nu_1, \ldots \nu_h$ are of the same order of magnitude as the numbers of molecules in ordinary bodies,

$$H - \eta_{gen} = \frac{\Omega + \mu_1\nu_1 \ldots + \mu_h\nu_h - \psi_{gen}}{\Theta}$$

$$= \frac{\Omega + \mu_1\nu_1' \ldots + \mu_h\nu_h' - \psi_{gen}'}{\Theta}$$

$$- \left(\frac{d^2\psi_{gen}}{d\nu_1^2}\right)' \frac{(\Delta\nu_1)^2}{2\Theta} - \left(\frac{d^2\psi_{gen}}{d\nu_1 d\nu_2}\right)' \frac{\Delta\nu_1\Delta\nu_2}{\Theta} \ldots - \left(\frac{d^2\psi_{gen}}{d\nu_h^2}\right)' \frac{(\Delta\nu_h)^2}{2\Theta}, \tag{553}$$

$$e^{H-\eta_{gen}} = e^C e^{-\left(\frac{d^2\psi_{gen}}{d\nu_1^2}\right)' \frac{(\Delta\nu_1)^2}{2\Theta} - \left(\frac{d^2\psi_{gen}}{d\nu_1 d\nu_2}\right)' \frac{\Delta\nu_1\Delta\nu_2}{\Theta} \ldots - \left(\frac{d^2\psi}{d\nu_h^2}\right)' \frac{(\Delta\nu_h)^2}{2\Theta}}, \tag{554}$$

where $\qquad C = \frac{\Omega + \mu_1\nu_1' \ldots + \mu_h\nu_h' - \psi_{gen}'}{\Theta}, \qquad (555)$

and $\qquad \Delta\nu_1 = \nu_1 - \nu_1', \qquad \Delta\nu_2 = \nu_2 - \nu_2', \qquad \text{etc.} \qquad (556)$

This is the probability of the system $(\nu_1 \ldots \nu_h)$. The probabilty that the values of $\nu_1, \ldots \nu_h$ lie within given limits is given by the multiple integral

* Strictly speaking, ψ_{gen} is not determined as function of $\nu_1, \ldots \nu_h$, except for integral values of these variables. Yet we may suppose it to be determined as a continuous function by any suitable process of interpolation.

$$\int\cdots\int e^{c} e^{-\left(\frac{d^2\psi_{\text{gen}}}{d\nu_1^2}\right)'\frac{(\Delta\nu_1)^2}{2\Theta}-\left(\frac{d^2\psi_{\text{gen}}}{d\nu_1 d\nu_2}\right)'\frac{\Delta\nu_1\Delta\nu_2}{\Theta}\cdots-\left(\frac{d^2\psi_{\text{gen}}}{d\nu_h^2}\right)'\frac{(\Delta\nu_h)^2}{2\Theta}} d\nu_1\ldots d\nu_h. \tag{557}$$

This shows that the distribution of the grand ensemble with respect to the values of $\nu_1, \ldots \nu_h$ follows the "law of errors" when $\nu_1', \ldots \nu_h'$ are very great. The value of this integral for the limits $\pm\infty$ should be unity. This gives

$$e^{c}\frac{(2\pi\Theta)^{\frac{h}{2}}}{D^{\frac{1}{2}}} = 1, \tag{558}$$

or

$$C = \tfrac{1}{2}\log D - \frac{h}{2}\log(2\pi\Theta), \tag{559}$$

where

$$D = \begin{vmatrix} \left(\frac{d^2\psi_{\text{gen}}}{d\nu_1^2}\right)' & \left(\frac{d^2\psi_{\text{gen}}}{d\nu_1 d\nu_2}\right)' & \cdots\cdots & \left(\frac{d^2\psi_{\text{gen}}}{d\nu_1 d\nu_h}\right)' \\ \left(\frac{d^2\psi_{\text{gen}}}{d\nu_2 d\nu_1}\right)' & \left(\frac{d^2\psi_{\text{gen}}}{d\nu_2^2}\right)' & \cdots\cdots & \left(\frac{d^2\psi_{\text{gen}}}{d\nu_2 d\nu_h}\right)' \\ \cdots & \cdots & \cdots & \cdots \\ \cdots & \cdots & \cdots & \cdots \\ \left(\frac{d^2\psi_{\text{gen}}}{d\nu_h d\nu_1}\right)' & \left(\frac{d^2\psi_{\text{gen}}}{d\nu_h d\nu_2}\right)' & \cdots\cdots & \left(\frac{d^2\psi_{\text{gen}}}{d\nu_h^2}\right)' \end{vmatrix} \tag{560}$$

that is,

$$D = \begin{vmatrix} \left(\frac{d\mu_1}{d\nu_1}\right)' & \left(\frac{d\mu_1}{d\nu_2}\right)' & \cdots\cdots & \left(\frac{d\mu_1}{d\nu_h}\right)' \\ \left(\frac{d\mu_2}{d\nu_1}\right)' & \left(\frac{d\mu_2}{d\nu_2}\right)' & \cdots\cdots & \left(\frac{d\mu_2}{d\nu_h}\right)' \\ \cdots & \cdots & \cdots & \cdots \\ \cdots & \cdots & \cdots & \cdots \\ \left(\frac{d\mu_h}{d\nu_1}\right)' & \left(\frac{d\mu_h}{d\nu_2}\right)' & \cdots\cdots & \left(\frac{d\mu_h}{d\nu_h}\right)' \end{vmatrix} \tag{561}$$

Now, by (553), we have for the first approximation

$$\overline{\text{H}} - \bar{\eta}_{\text{gen}} = C = \tfrac{1}{2}\log D - \frac{h}{2}\log(2\pi\Theta), \tag{562}$$

and if we divide by the constant K,* to reduce these quantities to the usual unit of entropy,

$$\frac{\overline{\text{H}} - \bar{\eta}_{\text{gen}}}{K} = \frac{\log D - h\log(2\pi\Theta)}{2K} \tag{563}$$

* See page 184–186.

This is evidently a negligible quantity, since K is of the same order of magnitude as the number of molecules in ordinary bodies. It is to be observed that $\bar{\eta}_{\text{gen}}$ is here the average in the grand ensemble, whereas the quantity which we wish to compare with $\overline{\text{H}}$ is the average in a petit ensemble. But as we have seen that in the case considered the grand ensemble would appear to human observation as a petit ensemble, this distinction may be neglected.

The differences therefore, in the case considered, between the quantities which may be represented by the notations *

$$\overline{\text{H}_{\text{gen}}}\rfloor_{\text{grand}}, \quad \overline{\eta_{\text{gen}}}\rfloor_{\text{grand}}, \quad \overline{\eta_{\text{gen}}}\rfloor_{\text{petit}}$$

are not sensible to human faculties. The difference

$$\overline{\eta_{\text{gen}}}\rfloor_{\text{petit}} - \overline{\eta_{\text{spec}}}\rfloor_{\text{petit}} = \lfloor\nu_1 \ldots \lfloor\nu_h,$$

and is therefore constant, so long as the numbers $\nu_1, \ldots \nu_h$ are constant. For constant values of these numbers, therefore, it is immaterial whether we use the average of η_{gen} or of η for entropy, since this only affects the arbitrary constant of integration which is added to entropy. But when the numbers $\nu_1, \ldots \nu_h$ are varied, it is no longer possible to use the index for specific phases. For the principle that the entropy of any body has an arbitrary additive constant is subject to limitation, when different quantities of the same substance are concerned. In this case, the constant being determined for one quantity of a substance, is thereby determined for all quantities of the same substance.

To fix our ideas, let us suppose that we have two identical fluid masses in contiguous chambers. The entropy of the whole is equal to the sum of the entropies of the parts, and double that of one part. Suppose a valve is now opened, making a communication between the chambers. We do not regard this as making any change in the entropy, although the masses of gas or liquid diffuse into one another, and although the same process of diffusion would increase the

* In this paragraph, for greater distinctness, $\overline{\text{H}_{\text{gen}}}\rfloor_{\text{grand}}$ and $\overline{\eta_{\text{spec}}}\rfloor_{\text{petit}}$ have been written for the quantities which elsewhere are denoted by $\overline{\text{H}}$ and $\bar{\eta}$.

entropy, if the masses of fluid were different. It is evident, therefore, that it is equilibrium with respect to generic phases, and not with respect to specific, with which we have to do in the evaluation of entropy, and therefore, that we must use the average of H or of η_{gen}, and not that of η, as the equivalent of entropy, except in the thermodynamics of bodies in which the number of molecules of the various kinds is constant.

Printed in the United States
By Bookmasters